Don't Be Fooled

In the debate leading up to the EU referendum in the United Kingdom the British politician Michael Gove declared that 'people in this country have had enough of experts'. In the 2016 Presidential campaign in the United States, Donald Trump was waged a war against the very idea of expertise. Yet if you are worried about your child's behaviour, don't know which laptop to buy, or just want to get fit, the answer is easy: ask an expert.

Where do we draw the line? Why do we appear to know more and more collectively, yet less and less individually? Has expertise painted itself into a corner? Can we defend both science and common sense?

In this engaging and much-needed book Jan Bransen explores these important questions and more. He argues that the rise of behavioural sciences has caused a sea change in the relationship between science and common sense. He shows how – as recently as the 1960s – common sense and science were allies in the battle against ignorance, but that since then populism and chauvinism have claimed common sense as their own. Bransen argues that common sense is a collection of interrelated skills that draw on both an automatic pilot and an investigative attitude where we ask ourselves the right questions. It is the very attitude of open-minded inquiry and questioning that Bransen believes we are at risk of losing in the face of an army of experts.

Drawing on fascinating examples such as language and communication, money, the imaginary world of Endoxa, domestic violence, and quality of life, *Don't be Fooled: A Philosophy of Common Sense* is a brilliant and wry defence of a skill that is a vital part of being human.

Jan Bransen is Professor of Philosophy of Behavioural Science at Radboud University in the Netherlands. He is the founder of *Philosophical Explorations* and has written scholarly work on practical identity, autonomy, narrative agency and love. Besides that, he publishes accessible books on the importance of cultivating a philosophical attitude to science, politics, media, mental health and modern life in general.

Don't Be Fooled

A Philosophy of Common Sense

Jan Bransen

Translated by Fulco Teunissen and Kate Kirwin,
Twelvetrees Translations

Routledge
Taylor & Francis Group

LONDON AND NEW YORK

First published 2013
Laat je niets wijsmaken. Over de macht van experts en de kracht van
gezond verstand
By Uitgeverij Klement, Zoetermeer.

First published in English 2017
By Routledge
2 Park Square, Milton Park, Abingdon, Oxon, OX14 4RN

Simultaneously published in the USA and Canada
by Routledge
711 Third Ave., New York City, NY. 10017

Routledge is an imprint of the Taylor & Francis Group, an informa business

British Library Cataloguing-in-Publication Data
A catalogue record for this book is available from the British Library

Library of Congress Cataloging-in-Publication Data
Names: Bransen, Jan, 1958– author.
Title: Don't be fooled : a philosophy of common sense / by Jan Bransen;
translated by Fulco Teunissen and Kate Kirwin, Twelvetrees Translations.
Other titles: Laat je niets wijsmaken. English
Description: 1 [edition]. | New York : Routledge, 2017. |
Includes bibliographical references and index.
Identifiers: LCCN 2016053633| ISBN 9781138716735 (hardback : alk. paper) |
ISBN 9781138716773 (pbk. : alk. paper) | ISBN 9781315189208 (e-book)
Subjects: LCSH: Common sense. | Expertise.
Classification: LCC B105.C457 B7313 2017 | DDC 001 – dc23
LC record available at https://lccn.loc.gov/2016053633

ISBN: 978-1-138-71673-5 (hbk)
ISBN: 978-1-138-71677-3 (pbk)
ISBN: 978-1-315-18920-8 (ebk)

Typeset in Joanna MT
by Florence Production Ltd, Stoodleigh, Devon, UK

Contents

3.   Understandable, efficient and good behaviour

4.   Trust and accommodation

5.   Dealing with grey

PART 2
Living with expertise   101

6.   Humaning today

7.   Waking up without science

8. Understanding without objectivity

9. Life satisfaction without policies

10. Responsibility without expert witnesses

Epilogue   182

Preface to the English edition

In the run-up to the Brexit vote, Michael Gove declared that 'people in this country have had enough of experts'. By saying this, the leader of 'Vote Leave' really touched one of the raw nerves of our current knowledge society, as paradoxically we feel enslaved rather than liberated by the enormous increase in expertise. Collectively we know more and more, while at the same time individually each one of us knows less and less. The number of experts is growing, but the domain of their expertise is diminishing. This is why even the best expert has to admit that he is actually a layperson in areas that are ever expanding. The more we know collectively, the more we are dependent on each other, and the smaller the number of subjects that we ourselves can discuss as experts. In the past, we used to be able to say that we knew about technology, or literature, or biology. But nowadays our expertise is limited to semiconductor testing, or South African poetry, or bacterial cell-wall structure.

Most of this expertise can be found on the Internet, and this phenomenon seems to suggest that anybody can be an expert on any given subject. But paradoxically, the wonderful, egalitarian availability of all this expertise only reinforces the ambivalent character of our knowledge society. Imagine you have a question about a certain subject: as a matter of course, you start searching the web and you end up with a substantial variety of detailed answers. How can you evaluate the differences between these answers? How can you determine which of the answers is the best one for your particular situation? Of course, if you sprain your ankle or have a nosebleed, it is handy that the Internet can tell you what to do. But suppose you have some inexplicable health problem and you visit a forum of experts by experience. This is not going to make you happy. Even if you are only a bit of a hypochondriac, you will probably end up self-diagnosing all kinds of lethal diseases. And all that is left for you now is to go to your GP to ask him[1] to confirm your diagnosis. Your search may have given you the impression that the Internet has made you a responsible and well-informed patient, but for your GP this will not be an unqualified pleasure, because he will realise full well that he can't really discuss your complaints with you at the same level of knowledge. This places a great demand on his communicative skills; indeed, not an area in which he is an expert. Of course, he may become an expert in communication, but he also has to sink his teeth into the latest financial software and health insurance guidelines, as well as keep tabs on any developments in pharmaceuticals, changes in his pension, his disability insurance and so on. Technically it is possible to follow all developments, but that is exactly why it is actually nigh on impossible. There is simply much too much expertise available for us to really master it all.

In this book I argue that expertise is slowly but surely affecting our common sense. Experts and laypeople are in deadlock, and this has brought about a situation in which both are forgetting to use their common sense. Because what is it that is really happening? There are plenty of experts for the tiniest aspects of our lives; as a result, when you are faced with some difficulty, it feels as if all sorts of experts are looking over your shoulder, experts who are much more knowledgeable than you. It is quite likely that this will undermine your confidence in your own independent thinking powers. After all, it is probably better to ask an expert. Is your child behaving in a way that raises a number of questions? Don't start pondering this yourself: ask a child psychologist! That is what he is there for; he has been trained to help in these sorts of situations. Can't you lose weight? Go and see a dietician. Want to stop smoking? Find yourself a coach and get some nicotine patches. Science has an answer for everything!

Something similar is happening to the experts. It is true that scientists are used to dealing with their own knowledge in a critical manner; they realise the provisional nature of their judgements and are aware of the necessary nuances and qualifications. But a layperson simply wants an answer: he simply wants to know what is what and he has no need for an expert who emphasises all the things that we do not know yet. A layperson desires the clarity found in headlines: 'Alcohol gene discovered'; 'Effective drug against ADHD'; 'Processed meats cause cancer'. In a competitive climate it becomes tempting for an expert to formulate his answers in a firm and strong manner, and leave his reticence at home. And there is more. We expect our experts to keep providing knowledge that is practical and useful in society. We encourage experts to restrict themselves to tried and tested models and methods, so that they can routinely produce confident answers.

The populist aversion that Gove tried to exploit in this predicament does not provide any answers, but is only a way in which the problem becomes apparent. The manner in which our knowledge society has developed a complete trust in modern science has led to an unfortunate situation of deadlock for both experts and laypeople, albeit in unforeseen and unintended ways. During the Enlightenment this total trust in science seemed like a great idea. After all, we had to shake off the irrationality of both capricious worldly powers and obscure religious powers. The Enlightenment philosophers argued that we had to build on our own intellect and our own rationality. What better way to do this than by building on science? And this is what happened, with impressive results! After all, in our daily lives we have become incredibly more healthy and wealthy in the past 200 years. Without any reservations this can be ascribed to the growth of scientific knowledge, a growth that we are all benefitting from.

But a change seems to be taking place. I argue in this book that the rise of the behavioural sciences has caused a change in the relationship between science and common sense. Only a short while ago, in the 1960s, common sense and science were allies in the struggle against political arbitrariness and religious superstition. But since then, this alliance has developed more and more into a struggle, or so it seems, between scientific expertise and our everyday common sense. It is as if our common sense is a quagmire of irrational, sentimental and chauvinist prejudice, of unfounded intuition that needs drastic correction by sound scientific expertise.

Populists have pounced on this conflict and like iconoclasts, they claim foolishly that they know best. As if our unfounded and uninformed gut feeling would be enough. I demonstrate in this book that this is an exceedingly one-sided and harmful view of our common sense. I will show you that our common sense is a collection of interrelated skills, whose use can be summarised in the following slogan: *Automatic pilot if possible and investigative attitude if necessary*. The automatic pilot is the product not only of our biology, but also of the social and normative environment in which we live. And the investigative attitude is in fact a skill that is easy to develop: asking *yourself* the right questions. It is this investigative attitude that we are slowly throwing away as a result of our dependence on expertise. After all, in a society that consists of laypeople and experts – experts who themselves are laypeople in all other domains of existence – it is the laypeople who ask the experts questions; as a result, hardly anybody asks *themselves* any questions anymore.

One of the characteristic deficiencies in today's populist movements is their inability to ask themselves a question, and this is true for the UK Independence Party, for Trump and the Tea Party in the US, for the Front National in France and for comparable movements in smaller countries such as Hungary, Finland, Austria and the Netherlands. Populists like to think in terms of 'us against them', with a clear and orderly division of good and bad, with the people who are on their side versus the people on the side of 'the oppressing elite'. Such populism has no place for ambiguity. Primitive emotions are rife, such as anger and fear in their raw, underdeveloped immediacy. If you think that such populism is the voice of our common sense, you have understood very little of this fascinating human ability, and this is what I will demonstrate in this book, which may be read as a critique of populism. But more than this, the book is a critique of the framing of our common sense as an ability that should be corrected by science.

Such a comprehensive, pretentious position deserves a careful and well-balanced analysis and argumentation. This is why this has become a full-length book rather than just a pamphlet. Our common sense is a wonderful skill, or rather a collection of interrelated skills, that deserves a detailed exposition. Part 1 of this book is devoted to this exposition, and Part 2 comprises a number of essays that can be read independently, essays in which I discuss a number of problems which are a consequence of the disturbed balance between common sense and expertise. I will show that this balance cannot be restored by an appeal to more expertise. I argue that it is in fact only our common sense that can help us redress the balance.

But before I invite you to the heart of the matter, I would like to explore the stage on which this misplaced struggle between common sense and expertise takes place. This is the sphere of our everyday language, the language that we need to tell a good story about our own lives, a story that gives us a foothold in our daily affairs. It is this sphere that I explore in the introduction to this book, in which I will confront you with the ways in which our everyday language both liberates us and keeps us captive in our attempts to build on our own common sense, our own rationality.

Note

1   I shall use "him" and "he" thoughout in a gender neutral way.

Introduction

How words liberate and captivate people

1. Autism

Teddy bear

She is standing at the top of the stairs, aged two and a half, and raises her hand holding her teddy bear.

'Teddy', she says.

Her father, half behind her, understands and takes the teddy bear. Satisfied, his daughter then grabs the handrail and carefully goes downstairs, step by step.

She doesn't need more than one single word to conjure up a complete scenario and to realise her goal.

'Listen here, Daddy. I would like to go downstairs and I am now big enough to do so on my own if I hold on to the handrail. But look, I am still carrying my teddy bear, and that makes it impossible of course; not only am I already big, but I'm also still little. So if you can take my teddy bear, I can go downstairs by myself. And then you can give me back my teddy bear downstairs, because you are big enough to go downstairs with a teddy bear in your hand. Is that alright?'

Dasein

Martin Heidegger, the famous German philosopher, is sitting at his old oak desk. He reads the passage he has just written and nods approvingly.

'In contrast, for our Dasein, this — that we understand Being, if only in an indefinite way — has the highest rank, insofar as in this, a power announces itself in which the very possibility of the essence of our Dasein is grounded. It is not one fact among others, but that which merits the highest worth according to its rank, provided that our Dasein, which is always a historical Dasein, does not remain a matter of indifference to us.

Yet even in order for Dasein to remain an indifferent being for us, we must understand Being. Without this understanding, we could not even say no to our Dasein.'

(Heidegger, 1953/2000, p. 116)

What you say is not always clearer if you have more words at your disposal.

But the more words you have, and the greater the care with which you pick your words, the more independent the message will become from its context, its sender and its recipient. In the Teddy bear scenario, one word is enough, at least if your father is a good listener and you can hand him your bear. But what if, as in the Dasein scenario, you need to say something about an abstraction that you can't point at and to a listener who doesn't know you and may have a completely different way of thinking? What if you want to say something to a person unknown to you about the behaviour of someone who this person doesn't understand? In such cases you need an extensive vocabulary, a vocabulary that is objective and that is rooted in reality rather than in your communicative skills, a language that belongs neither to you nor to the other person, a language spoken by the world in itself.

Such an independent language with objective foundations is an old ideal – an emancipatory ideal from the Enlightenment. If words don't belong to anybody in particular, this gives us the greatest chance that no subjective distortions will surreptitiously sneak into our utterances. At least, this is one of the presuppositions that have come along with the idea that modern science has a liberating and democratic effect because it provides us with a wonderful human ability: critical thinking. In this book I investigate the plausibility of this presupposition and pose questions about the role of expertise and common sense in the development and use of a language necessary for people to understand both themselves and each other. I will start off this investigation with a third scenario, a scenario that was unthinkable 20 to 30 years ago; not only have I witnessed this myself in the past year, but I have also heard similar stories from other people. Slowly but surely scenarios like this are becoming the norm.

Michael

We are sitting in a circle. Sixteen people who do not know each other, but who have all joined a bookbinding course, which we will be following together in the coming weeks. It is the first night of the course. The teacher, who will turn out to be fine, has not investigated the variety of modern approaches that are available to get to know each other a little bit in a more or less surprising manner. Instead, he has opted for the traditional round of introductions, in which everybody shares information about themselves and their expectations regarding the course. Michael is number four.

'Hi, my name is Michael. I am autistic. This is not something that people can easily see about me, and that's why I'm telling you.'

Everybody is listening and nodding understandingly. This is the way things go nowadays. This is the way you introduce yourself.

2. Zoon logikon: talking animals

The behavioural sciences, a relatively recent invention, have had an intriguing effect on human existence. Of course, 150 years ago there were also people who would now be regarded from a neurological or behavioural point of view as people with an autistic disorder. But that was not what they were called then. Nor was it how this was experienced, either by the people themselves or by their environment. In those days, there was no way for people to diagnose themselves as autistic. And it was impossible to deal with relatives, friends and acquaintances in a way that could be characterised and summarised by the label 'autistic'. Nowadays, however, this is a respectable way to find your place in society (Cf. Hacking, 2002). Michael is autistic. Clear. You know immediately who you are dealing with. At least, if you have enough information, if you know enough about behavioural disorders, and if Michael gives you some additional information about his specific subtype within the autism spectrum.

People are talking animals. They use their voices when they introduce themselves. 'My name is Michael.' Their physical appearance is apparently not sufficient to know who they are. They say their name and add, if there is enough space and time and if it is expected, a label, a description or a story. For example, they can say what they do for a living, where they are from, what they like, or anything else that is characteristic of them. 'I am autistic.' If you get to talk to them in greater detail, you will notice that they draw much more on our common language. They can tell you complete stories about themselves and the most interesting aspect of these stories is that they are not superficial stories, similar to commercial posters or window dressing, but that they are stories that are fundamental to how a person experiences himself, understands himself, is himself.

I am not saying that these stories are true or untrue; I am not nearly ready for such difficult issues. Nor is that my point. At the beginning of your bookbinding course you can easily say that you are fascinated by old craft, that you love books, reading, culture and tradition, and that you adore the musty smell of old books that you always sniff attentively when you open a tome. You may even start believing this story yourself, even if you know that the real reason you are using such an introduction is that you have joined the course because you are looking for a relationship and that you believe that these supposed confessions will make you more attractive to women. When you introduce yourself you add verbal layers of the person you are, or think you are, to your bodily presence.

This is no deception. But of course you can fool the other, or fool yourself. Of course you can pretend to be much more interesting than you really are and you can theatrically pick up an old volume and sniff the cover. And since you have begun to fantasise, you may well continue to imagine how the ladies at the bookbinding course start fidgeting, how they sigh quietly and decide to approach you during the break to confess they too adore that musty smell. If you're inclined to fool yourself, I don't mind fanning the flame.

Nevertheless, people who introduce themselves are usually sincere. With a fancy word you could call these introductions personal identity management. (Bransen, 2008) This in no way means that you are pretending to be something you are not. You are presenting yourself, trying to make visible what is not immediately visible, in this case yourself. For this you need an increasing number of words, something you notice when you get older. You can use these words like you use a torch, to illuminate dark holes and corners. The words you use to introduce yourself have a similar effect to the torch in the dark: the ray of light not only illuminates the item it focuses on, but also deepens the darkness around it. What is outside the centre of attention becomes darker and less visible.

Even though truth and truthfulness are important in this matter, in the case of self-presentation and self-interpretation it is not immediately clear what is true and truthful. If you introduce yourself, you do not give an objective account of what you know about yourself. It is not like reading a thermometer or giving directions. If you introduce yourself, you are creating something, putting something into words; it is expressive rather than a neutral exchange of information (Cf. Taylor, 1985).

And exactly because it is expressive, because it is a matter of putting something into words, of finding the right words, I have elaborated on what would happen if Michael were to introduce himself by saying that he is autistic. People – talking animals, zoon logikon as the ancient Greeks put it – not only live in the world and are not only a physical body, but they also live in their stories and are also their voice. I am especially interested in the fact that these stories are built with and around the words that people have been given by their common language. This language belongs to everybody in general and to nobody in particular. This language is alive, with us and among us. But even though this language belongs to everybody, that doesn't mean that everybody has the same influence on the language. Some groups of speakers have more influence on the development of a language because they are seen as experts in their fields.[1] My gardener taught me how to use the words 'goutweed' and 'horsetail'. My GP told me what is meant by an irritable bowel and by Crohn's disease. And the TV sports commentator explained the difference between 'man-on-man coverage' and 'zone coverage'.

If we want to examine human self-interpretation, what people say about themselves, the concepts they have learned to use to discuss their own behaviour, their motives, their attitudes, their outlook on life in general and their own life in particular – if we want to examine all that, Western culture has seen a long procession of experts. This procession is led by ancient Greek philosophers like Socrates and Aristotle, closely followed by physicians like Hippocrates and Galenus, church fathers like St Augustine and St Jerome, theologians like Thomas Aquinas and Martin Luther, humanists like Erasmus and Spinoza, physicists like Newton and Einstein, authors like Shakespeare and Goethe, political thinkers like Machiavelli and Montesquieu, and modern philosophers like Kant and Wittgenstein. However, since the advent of empirical science, since psychology disentangled itself from philosophy, and since healthcare started reorganising itself along lines of economic feasibility, a new type of expert has appeared: the behavioural scientist, who knows his way around the most up-to-date version of the DSM, the Diagnostic Statistical Manual of Mental Disorders. Since then, a new well-founded vocabulary has become available for if we want to (or have to) introduce ourselves.

This new vocabulary holds a promise. In the seventeenth century, Blaise Pascal explained the vulnerability of human existence as a lack of self-knowledge. He stated that even though mankind is capable of knowing all about nature around us, this will eventually lead us nowhere, as we do not understand our own human nature. However, with the advent of the modern behavioural sciences a vocabulary has become available with which we can accurately and truthfully tell the story about ourselves. 'I am Michael. I am autistic.' Here is the promise, analogous to the toddler standing at the top of the stairs, lucky enough to have a father who understands his daughter so that she needs to use only one word. It is the same with Michael:

— Listen people. It is not so easy for me to understand the processes going on between and among you. All I know is that I have a different mental make-up than you. I like to be on my own and I don't need to have so much contact with other people. I'd rather be home with my stamp collection. I have all my stamps neatly pasted in my album, exactly in the correct place. I am never sure what other people expect of me. I have a doctor, or actually a psychologist. A therapist. She says that it is good for me to see other people. She gave me the flyer for this course. She said that this may help me to have better interaction with my grannie, the neighbours and the people at my work. She said that it may help if I tell you what is the matter with me. Because you will understand that I am autistic. So. Okay?

3. Independent of context and audience

Cake

I am standing at the counter, and I bend over a little, keeping my head up so that I can see better.
— That one there. No, no, not that one, this one here. Yep, that one.
— Ah, you mean the Triple Layer Nutty Cappuccino Cake. How many slices would you like?
— Three, please.

Sometimes I overdo it and hold up three fingers. Quite easy, talking with your hands and feet. Of course you would learn the names of all the different cakes soon enough if you worked at a patisserie, as that would only be practical. After all, superficially it is all only a matter of using your brain economically, being slightly lazy. The next time I buy cake will be months from now, and by that time I will certainly have forgotten the name of that scrumptious little cake.
— Yep, that one there. The one I had last time.
The same thing happens at the fishmonger's. Or at the garage. Or in your own garden, when you have finally decided to bring in a professional gardener, to whom you point out the weeds in your garden. 'Goutweed, horsetail, shepherd's purse, knotgrass, bellbind.' Such silly names! They go in one ear and out the other. After all, they are only names, and they are only weeds. It's actually rather pretty, bellbind.
It is more serious at the GP's. She also uses classifications. And she also looks for the right word, asks questions, presses on your stomach, uses the stethoscope to listen to your heart and lungs, uses a little light to look into your eyes – all that to reach the right diagnosis. But the GP doesn't use just any old terms. They are all related to a certain therapy and it is sometimes a matter of life and death. But even for someone of a mildly hypochondriac disposition, 'sometimes' may be often enough, and scary enough. Life and death, that doesn't bear thinking about. That's why it is so fortunate that the GP is an expert. She doesn't just name the

symptoms. She is a representative of medical science, who uses a universal language to reach an objective diagnosis, who studies the underlying mechanisms at micro- and nowadays even at nano-level, and who systematically evaluates her treatment, so that she can take pride in impressive successes and can make use of evidence-based interventions. Welcome to the twenty-first century!

One crucial aspect of medical vocabulary and the medical profession is that the diagnosis, treatment and theory-building are independent of your own GP's individual characteristics. At least, if you have a good GP. Medical science is objective, and its language is anchored in reality rather than in our ways of interacting. For example, herpes zoster is shingles, and it is the same the world over. If you have that particular disease, any doctor can make the diagnosis and prescribe treatment. Even if you are on the other side of the world and you are suddenly confronted with the symptoms of this disease, and you can only explain what you are feeling by using your hands and feet, the doctor who you visit there and who only speaks Vietnamese will be able to make the diagnosis. He will give you a prescription that you can probably not read, but fortunately the chemist that you will visit later is also an expert; he can easily read the illegible scrawl and will give you the correct antibiotics. It is a relief that even the final details can be communicated using your hands and feet. Three pills a day. Three? You hold up three fingers. He nods. Good.

The existence of an objective vocabulary, a vocabulary in which a message can be formulated such that its meaning is completely independent of the context, the speaker and the recipient, is an invaluable cultural achievement. An objective vocabulary that is actually embedded in reality is an unparalleled ideal that has captured the imagination for over 2,500 years. It is uncertain whether modern science has been able to reach or even to approach this ideal. However, as a result of its tremendous technological and medical successes, at first sight science seems to have been proven right.

Critical modesty should be part of this success. But our growing knowledge comes with all kinds of snags. And of course it is much more difficult to have a well-defined meaning, independent of context and recipient, for a concept such as AUTISM than for concepts such as THREE and HERPES ZOSTER. However, it is not impossible. At least, there doesn't seem to be any fundamental reason why science would not be able to map human behaviour. Just take a look at the major advances made in mapping physical, biochemical and neurological reality. Especially because modern science knows the enormous challenges it faces, it can keep its ambitions high and offer mankind a methodological framework for the development of a vocabulary that can guarantee objective independence.[2]

This vocabulary needs gatekeepers, experts who guard the meaning of the terms used. Not everybody understands HERPES ZOSTER well enough to define this concept; physicians do. Not everybody understands GOUTWEED well enough to define this concept; biologists do. The following is also true: not everybody understands AUTISM well enough to define the concept, but scientists who know the DSM do. These gatekeepers, these experts, have provided mankind with an objective language, a language in which the terms mean what they mean, independent of context, sender and recipient.

Even if you have insufficient expertise to define certain scientific concepts, you can still learn how to use them. This is satisfying and important, as it also gives non-experts the opportunity to find their way in a world that is continually becoming more complex. If you have had herpes zoster a number of times, you will have learned to recognise the symptoms and you will be able to have a discussion about this disease with your doctor. You know when you have shingles and also when it is definitely something else. Similarly, you can become an expert in weeds when you notice that goutweed keeps on raising its ugly head. After a while you recognise that irritating weed from miles away. The same goes for Michael and for us, his twenty-first-century fellow humans. By now we have learned to use the term 'autism'. We have heard a great deal about it and come across it often. We have learned to recognise autism as one way of being a human being. To us autism is just a thing that some people have. It is beginning to become normal.

4. Missing information subroutine

Suppose you are in a particular situation and you are not quite sure how to go on. For example, you are in town, looking for a shop, but you are not quite sure where this shop is and you are already late. Or you're cooking dinner, trying out a new recipe, following the instructions but it doesn't look as if it is going to turn out the way you expected. Or you're reading a book, for example this one, and you are losing your feet, for example like now. Where is this all heading? You can't really follow the reasoning anymore. What are you going to do?

The sensible thing would be to give yourself a moment to think, to take a step back and observe the situation from a distance. How did you get to be in this situation? What is the logical structure of this scenario? And does that help you to determine with some clarity what the next logical step would be? So when we consider the book scenario, let's briefly look back together, from a little distance. In a nutshell, these are the steps that we've made so far (or at least the steps I have suggested):

- This book is about the role that expertise plays in our daily lives.
- This role is much to do with language.
- Experts are the developers and gatekeepers of different vocabularies with a presupposed objective validity.
- At present, behavioural scientists are the experts who guard the vocabulary that we apparently need to use in order to understand ourselves.
- Behavioural scientists strive for a language that is independent of context, sender and recipient, a language which is primarily anchored in reality rather than in communicative interaction.
- Behavioural scientists have not only introduced new concepts such as AUTISM and OBESITY, but have also given a respectable scientific content to well-known concepts such as ADOLESCENCE and EDUCATION.
- As a result of this language, our lives are increasingly controlled by experts.

Before elaborating on these thoughts, I will first discuss a way of thinking that has become dominant: using the missing information subroutine as a starting point. The concept subroutine comes from ICT. My ICT knowledge stems from when I was a student of philosophy and I bought a Commodore 64, the first computer for consumers, with which you could perform miracles. The number 64 referred to its working memory, which was its only memory. It had no hard disk, no disk drive and no monitor either. It did have a working memory of no less than 64 kilobytes. Quite, kilobytes! That was all.

Anyway, I started programming away and soon became familiar with the concept of the subroutine. It works as follows. A computer goes through its program, which is a list of lines. Sometimes it can only go from one line to the other, for example from 13 to 14, if it knows the value of a certain variable. If it doesn't know this value, line 13 starts a subroutine, a small program whose only aim is to discover the value of that variable. I'll give you a simple example. I wrote a small program so that the computer could write an invitation to a party. The invitation started with 'Dear'. The next thing was to insert a name, everything automatically, so this would be the next name on a list of friends. To know which name should be inserted, the program started a subroutine, an additional program that went through a saved list of friends and checked which name was the last to have been used. The next name was then the one to be inserted in the invitation. The subroutine then passed on this name to the program and the greeting to the invitation could be written: 'Dear Elisabeth.'

In the same way, the missing information subroutine is a subroutine. You are in town, looking for a shop, but you are not quite sure where this shop is and you are already late. Walking around at random won't be productive. Neither will panicking or being angry with yourself. Of course you could ask somebody the way, but that is not the way a man does things. He can find his own solution, just by thinking for a moment. Sometimes you have no choice, for example if you are the first in a certain scenario, or if you are a pioneer, which we all are at some point, especially if it is about the direction of our own life. If you're stuck, if you don't quite know what to do, the obvious next step is to interrupt the scenario, take a step back and think. Start your missing information subroutine. If you can't find that shop in town, the subroutine will conjure up a mental map of the city centre. Calmly you picture the way the city centre is structured. You use this mental map to find the location of the shop, your own present location and the route from here to there. If this subroutine finds a result, like the route to the shop, it feeds it back to you, the actor in the original scenario, so that you can use this information to continue the rest of the program. Using the route you have just obtained, you can continue on your way, sensible, determined and informed.

This is the way people do things. And they have been doing it like this for centuries. Throughout evolution, people have trained themselves to work with missing information subroutines. It must have started in prehistoric times. No food to be found? Think. It may be a good idea to move on and live a nomadic existence. Winters that are too cold? Think. How about living in a cave and wrapping yourself in a bearskin? Feeling ill? Think. You may be able to find out what the cause is and make it go away. You want to know how far that ship is from the shore? Think. Pythagoras' theorem may come in handy.

The ancient Greeks turned this into a systematic way of thinking. They gave us the idea of a theoretical perspective, a disinterested standpoint, a withdrawal from the urgency of everyday practical issues so that you can calmly think things over. They also gave us the seeds of a societal arrangement in which some people, the experts, are especially concerned with thinking and who provide the society at large with the knowledge necessary to avoid getting stuck in practical scenarios. Plato founded the Academy, a place where problems were studied and solved. His example has been emulated by our present-day universities, who have slowly become skilled in a modern variety of thinking: going through a missing information subroutine.

Together, these two ideas have taken off in our modern knowledge society: first, the objective theoretical perspective, and second, the university as the societal arrangement that takes care of the missing information subroutine. I hope you are aware that I am allowing myself an absolute simplification here, simply to provide you with a powerful image of a special type of labour division. On the one hand there are people, laypeople, who sometimes need information, and on the other hand there are experts, scientists, who work at universities and who are specialised in the missing information subroutine for a small area. The area of expertise for these scientists is becoming increasingly smaller, as science is becoming more and more specialised. To a great extent, every expert is also a layperson. That is why we are more and more interdependent, which could be seen as a positive development. A top economist knows as little about herpes zoster as I do, and similarly a medical specialist does not know more about refinancing budget deficits than me. And so on.

The promise of an objective language finds its origin in the idea of the university as a collectively organised subroutine. For example, suppose that someone, a layperson, is not sure how to continue in a certain situation. These days, there is no need for such a person to ponder long about a solution. Most problems are too complicated anyway. As lay-people, we are unable to solve these problems simply by thinking. To find a solution we need a vocabulary that we may be able to use a little bit, but that we can't really fathom sufficiently. We need to leave that to the experts, and we are fortunate enough that we can leave it to the experts. If we don't know how to proceed, we can turn to the university. We can ask someone at the university, and in our advanced knowledge society there is a good chance that we will find a scientist who has just completed a thorough study of the scenario in which we got stuck. With a little luck we can soon continue our course in this tricky scenario, taken by the hand by the new knowledge provided by science.

Michael has autism. This is not easy, but nowadays, as a result of the enormous developments in behavioural science, it is much easier for people with autism. And thanks to science it has become much easier for us, his fellow human beings, to deal with people like Michael. That is a positive development. At least that is what you might think once you realise how hard scientists have been trying to help us, to map a complex disorder such as autism even better, and to develop and test better treatment options.

But this development is not only positive. We also lose something if the concepts we use to understand ourselves are becoming so complex and technical that we cannot understand these concepts ourselves. It is like we have to hand over parts of our own lives to other people. It is as if we really cannot answer some questions about ourselves, or to be more precise, as if we can't even ask the right questions anymore.

And that is a process that needs to be considered more closely.

5. An investigative attitude

When people think, it doesn't necessarily mean that they are in a missing information subroutine. Sometimes it is amazement that knocks us out of kilter, as the ancient Greeks put it. For a moment you just don't know. Something affects you, requires a reaction, or even just a little bit of understanding. In such situations people also tend to take a step back, have a close look at the situation, if only because they have no idea yet what the information is that they are missing. For example in the following three situations:

Out of kilter

For no apparent reason, your 9-year-old son suddenly explodes, angry, sad, helpless, upset. He wants nothing to do with you, is angry especially with you, but then all of a sudden breaks down on your lap and is a small, vulnerable child again.

After her final exams, your girlfriend is going to backpack around the world for 9 months. You don't like this at all and say that if she really goes, you will break up with her. You regret it the moment you say it.

Your neighbour is 62 years old and looks after her mother, who is infirm. Day in day out she is available for her mother, but now she's sitting on your settee, with a blank look in her eyes and has no idea how to go on.

The ancient Greeks made no distinction between science and philosophy. In their work there is no clear and significant difference between the demand for missing information and the demand for understanding. It is important to emphasise this. And it may be important to also emphasise the fact that the ancient Greeks had no computers and therefore could not see the computer as a metaphor for our mind. Their image of a thinking human being was not linked to the image of a system that processes information algorithmically. They did not know the idea of a subroutine, and they did not think that if you ask a question you are asking for missing information, for facts that you require in order to continue in a certain scenario. They continuously asked questions simply because they were wondering about all kinds of things. For them, amazement was a good occasion to take distance, to see things from an objective perspective, and to consider an issue theoretically. But for the ancient Greeks this did not necessarily mean that they needed missing information in order to continue.

Let me be as clear as possible about what I mean. The missing information subroutine that I described in the previous section is a small program, a strategically efficient way of gathering the missing information needed to carry out the main program. For example, you are writing invitations and you need the next name in your list of friends, you need milk and yoghurt, so before you can go to the checkout you need to find out where the dairy section is in this supermarket, or you are doing up your garden and you need to know which of the plants in your garden are weeds. The nature of these scenarios is always the same. You find yourself stuck because you lack certain relevant information, which is why you switch to the missing information subroutine, and in these cases this subroutine does not need to be more than a simple strategic program. It provides the missing information that gives you the opportunity to continue in your scenario.

However, the scenarios regarding the upset child, the backpacking girlfriend and the neighbour with emotional exhaustion ask a different kind of question. In these situations there's also something you do not know, but the similarity ends there, as it is not clear in these scenarios whether there is any missing information. That is certainly possible, but does not need to be the case. There is no completely obvious interpretation of your situation in these three scenarios that makes it clear to you that you're missing information and what this missing information is. Something else is the matter in these situations: you first need to contemplate the aspect that causes you to get stuck, that confronts you with your inability to simply continue. What you need here is an investigative attitude. And if this investigative attitude helps you to understand that you are missing certain crucial information, and you know how to obtain this information, then starting the missing information subroutine is the best and most obvious next step.

As I said earlier, the ancient Greeks made no distinction between philosophy and science, and as a result, for them amazement was always an occasion to adopt an investigative attitude. This investigative attitude can be seen as the anteroom from which a possible missing information subroutine can be started as soon as it is clear that there is indeed some information missing and we know where this information can be found. The anteroom is where the question is first analysed. If the question is mono-interpretable, you don't have to stay in the anteroom with an investigative attitude. You can simply continue to the missing information subroutine. But if you are wondering what to think about yourself and your ridiculous threat to your girlfriend, you do need this investigative attitude, as in this case it is not at all clear if what you are dealing with is only a lack of information.

In our modern knowledge society in which the missing information subroutine is taken care of by experts – empirical scientists at universities and other specialists in collecting missing information – we have more or less lost sight of the anteroom with its investigative attitude. In a nutshell, my diagnosis of this issue is the following.

First of all, if you get stuck in a certain situation nowadays, you may be lacking expertise and therefore you turn to an expert or to science. If you are wondering why you would say such a nasty thing to your girlfriend, you can go to a psychologist. It is probably something to do with jealousy, with stress, with hormones, with insufficient healthy attachment, and perhaps even with a temporary slump in your confidence as a result of the stage in which your adolescent brain finds itself. Nowadays you immediately turn to science because everything is so complex, which is why you do not need an investigative attitude anymore. Your anteroom has become science's waiting room. When you get stuck, you immediately go into your missing information subroutine. Without thinking, you assume that this is a matter of missing information, in fact of missing scientific information. And without thinking you know where you can get the answers: from science.

Second, experts at universities have also forgotten to adopt an investigative attitude. This may seem surprising, but isn't really. In fact scientists have become skilled in the experimental method. They are excellent experts in the field of operationalised questions, and they can empirically test their hypothetical answers in their laboratories. Because they are so well-trained in the methods that they use, they have become especially good at reformulating questions in such a way that these questions can be used in experiments. It is not a sense of wonder that drives these scientists but the set-up of their labs. If I may exaggerate a little, they are not concerned with the question but with the answers that can be produced in their laboratories. Or, to use a metaphor, simply because they have a hammer, they tend to misinterpret a screw for a nail. This is why experts nowadays also automatically go into the missing information subroutine, thinking that every question is a request for missing information.

Third, finding yourself is not a subject for experts. In a way this is correct. After all, you are the only expert regarding your own orientation. That is all right, as far as it goes. There are plenty of self-help books. But this is exactly where something extraordinary, interesting and even disquieting happens, something that leaves you empty-handed, notwithstanding our highly developed knowledge society. Because what is the language that you can use to discuss your own orientation? Of course, the language of the behavioural sciences, a language, however, of which you haven't yet mastered the finer details. To understand yourself, you need the experts who have withdrawn from daily life because they need distance, because only when they have distanced themselves from daily life can they take care of the collective missing information subroutine. To come back to the hammer metaphor: science likes to offer you its hammer, its expertise. It just doesn't realise how much work is involved for the hand holding the hammer. Something to ponder on.

6. Philosophy's anteroom

These days, philosophy's anteroom is rather deserted. True, there is a lot of interest in philosophy. Magazines on philosophy sell well and there have been Nights of Philosophy in Paris, London, Berlin, Amsterdam and New York. Philosophers are on the radio and on TV, they give TED talks, write blogs and post videos on YouTube that get over a million views. What more could you want?

I would like a bit more. I would like to gather everybody together in that anteroom of philosophy. And then I would like to see that investigative attitude, in each and every one of us. Because I do not like shows that are not about thinking but about looking at thinkers who use complicated words to show off about the number of books they have read. I don't like thinkers; I like thinking.

Nor do I like this self-inflicted dependence. I would like us to stop automatically relying on experts. I would like us to stop automatically thinking that we are missing information. I would like us to start thinking for ourselves. I would like us to take a good look at the questions first, the questions that cause us to become stuck in our daily lives, the questions that deserve attention. These are the questions that do not have a quick or easy solution; they need attention and need to be understood – as questions. They require understanding rather than some missing information.

In this book I take up arms against the advance of the experts and against the growing dominance in our daily lives of scientific expertise. I wish to emphasise that I am not pleading for an anti-intellectual or anti-scientific attitude. I deeply love thinking, using your brain; I like to do so systematically, critically, verifiably and methodically. I have nothing against our intellect, or against intellectuals. I have nothing against science, or against scientists.

But I do have something against societal arrangements that systematically reward our lazy brain. I do have something against waiting in the wings as a matter of course, because we think the answer has to come from experts. I do have something against mindlessly moving into the missing information subroutine as soon as something puzzles us. I do have something against expertise, at least against expertise if it is related to our daily lives and if it is based on the presupposition that there is not a gradual but a categorical difference between common sense and knowledge based on scientific research. I have something against the unnatural division between on the one hand daily practice in which people gain experience – in other words, knowhow – and on the other hand research laboratories in which experts obtain statistically significant results with which they justify their evidence-based interventions.

I have something against neglecting and distrusting common sense, as it is precisely this common sense that you need as soon as you want to appeal to scientific knowledge. Moreover, it is exactly this common sense that you use as a scientist in all your activities in the laboratory, in discussions, in theoretical reflections and when giving advice. And it is also this common sense that you need if you want to be able to do anything at all with the various scientific results that are flooding our daily lives.

I like the image of a scientist who does his best to fill our toolbox with the most fantastic attributes, instruments and tools. It is so much more positive than that rather ominous image of the scientist who thinks that he can do everything with just a hammer, and consequently mistakes a screw for a nail. But even in the image of the well-furnished toolbox, the most important element is missing: the hand. It is their hand, our hand, your own hand that is needed to do anything at all with all these wonderful tools. If something is wrong with your car, your toilet or your mobile phone, you are more than happy to have it repaired by somebody else, a skilled worker with the right toolbox. But if there is something wrong with your life, your attitude or your behaviour, you can't leave that to somebody else. At this point a toolbox is a welcome help. But then you really need to know how to use the tools. You want to learn how to use your hand, the hand that you have always had, that has been given to you by nature, with which you have done everything up to now and with which you will be doing everything in the future. After all, if you can't use your own hand, all the tools that the experts offer you will be useless.

This hand symbolises your common sense. It also represents the common sense of all experts, of all specialists. Without their common sense they wouldn't have got far, they wouldn't have even started. And all the knowledge and expertise that they have gained are only extensions of that common sense.[3] There doesn't have to be a contradiction. There is no conflict between scientific expertise and common sense. However, there is a gap unfortunately. A wide gap.

I don't think anybody will be surprised at this point that this gap can only be bridged by common sense. You can have as many tools as you like, but at the beginning of the chain there will always need to be a hand that operates the first machine, that knows what it has to do and knows why, for whom and to what end. It is this common sense, our common sense, that I will defend in this book.

This book has two parts. In the first part I explain what I regard as common sense, by following a group of people on a fictitious island who cannot fall back on the advice of experts. They will have to solve their problems themselves, using common sense. The tone in this first part is constructive: It is a story about the different aspects of common sense. The second part has a totally different tone. In this part, I try to create turbulence, encouraging the reader to think. By exploring a number of presuppositions related to the scientific orientation that nowadays dominates our daily lives, I try to show how questionable these presuppositions are. Both parts serve the same goal. The first part shows you what common sense actually is, whereas the second part shows you what you can gain if you use your common sense.

Notes

1   The nature and the implications of this social fact have long played an important role in Foucault's writings. See Foucault (1996/2002).
2   Bernard Williams' study of Descartes' philosophy brilliantly describes this modern ambition. See Williams (1978).
3   Quine is famous for emphatically supporting the growth of science. But in doing so he emphasised that disavowing the core of common sense is "a pompous confusion": "Science is not a substitute for commonsense, but an extension of it." Quine (1957), p. 2.

Part 1

Living with common sense

1 Humaning on Endoxa

1. A new verb

Creatively interpreting Aristotle, the Dutch theologian Harry Kuitert came up with the wonderful idea of considering the word 'humaning' a verb.[1] I human, you human, he humans, we are humaning. The brilliant aspect of this invention is that it emphasises that being human is a matter of *doing*. The fact that we are people means that we act like people, that we are *humaning*. Moreover, this invention provides the opportunity to present an interesting aspect of human existence, the fact that we can never evaluate our existence neutrally or objectively, as a more or less objective characteristic. What I mean is this. It sounds horrible and even wrong to talk about a good human being or a bad human being. But if you use the word 'human' as a verb, you can talk about 'humaning well' or 'humaning badly', as a grammatically correct fact that doesn't need to have negative associations. You can compare it with the didactically sound reprimand, "I don't disapprove of you, but I disapprove of your behaviour." Just like human beings can talk, sing, play football, and hold meetings, we can also human. And just as not everybody is equally good at singing or playing football, not everybody is equally good at humaning.

If I use my creaky voice to produce unintelligible sounds in which there is hardly any difference in pitch and no flowing connections between these sounds, you may conclude that I am *trying* to sing but am actually doing this so badly that you can't really call it singing at all. From a distance, singing should at least vaguely resemble what Pavarotti, Susan Boyle and Taylor Swift do. The same is true for playing football: it should at least have something to do with a ball that you move around with your feet, and at the highest level it should be similar to what Wayne Rooney and Lionel Messi do. This means that when you play football, you need to fulfil certain criteria. On the one hand these are criteria that determine whether what you are doing counts as an example of playing football (these are called constituent criteria), and on the other hand, whether what you do is a qualitatively good example of playing football. These are qualitative criteria, which are mostly implicit and difficult to formulate clearly and unambiguously. Of course, there are similar criteria for talking, holding meetings and so on.

If humaning is a verb, there will be similar criteria. If humaning becomes a matter of doing rather than being, this leads to interesting possibilities regarding the constituent criteria. The fact that dogs, dolphins, apes and robots are not specimens of the human species does not necessarily mean that they can't human. It may well be possible that they really can't do it, or that it only resembles humaning a little bit, but that won't be because they don't belong to *Homo sapiens*. This provides new and fresh opportunities to look for a decisive difference between either humaning very badly or not humaning at all. A snail won't be humaning, as it is not interested in humaning, just as an alarm clock or a standard lamp is not interested in humaning. But you can imagine that a dog, which plays an important part in the life of a family, sometimes shows behaviour that you could call humaning, or at least behaviour that is difficult to distinguish from the behaviour of the children or even the parents in the family. Conversely, you could imagine that you wouldn't want to call a human being's behaviour humaning if he kills a 'volunteer', cuts him into pieces and eats him with a glass of claret.

More intriguing, and much more interesting for my argumentation, are the qualitative criteria, the criteria that make the difference between somebody who is good at humaning and somebody who is bad at humaning. Do such criteria exist? As we have seen, they exist for other verbs, but in the case of humaning, such criteria are an extremely sensitive matter. Of course, not every example of humaning is an example of good humaning, but these are murky waters and if you are not careful, you may be reproached for having politically or morally reprehensible ideas. You could get away with criticising Hitler and Hannibal Lecter: there will not be many people who regard their actions as examples of good humaning. And also on a smaller scale there are several relatively uncomplicated criteria that you can discuss without accusations of being overly moralistic. We teach our children all sorts of different things, for example how to use the toilet, how to read and write, that they should take other people into account and that they should be honest and patient. Of course, there are more of these uncomplicated virtues. The implicit idea is that if you want to human well, these are the things you should do: go to the toilet in time, refrain from fighting, give other people some room too, be honest and so on.

You may have noticed from the care with which I have picked my words and the relative neutrality of my examples that I am well aware that our society is no longer a place for moralistic leaders. Half a century ago this was completely different. In those days it was the vicar or the priest who knew exactly what the acceptable qualitative criteria were for the verb humaning. And it wasn't just the vicar or the priest: the elevation of the people was also an important idea in the socialist workers' movement. It was absolutely clear that there were qualitative criteria for people, which combined well with a stratified classification of society. After all, some people are better at humaning than others. Of course, this situation was not immutable. Especially in the years that would culminate in the events of 1968, it was an *initial* stratification, around a historically developed inequality that we would be able to leave behind, simply by developing and educating ourselves. Humaning is not only something that we do naturally, but also something that we can train, that we can become better at, day by day, simply by having the opportunity to flourish as human beings.

At present, there are two interesting developments regarding the qualitative criteria of humaning as a verb. The first development is related to the strong tendency towards individuality and democracy that is characteristic of current society. The idea is that humaning has a different meaning for each of us, because it is mostly a question of becoming yourself, being yourself, developing your own authentic style, and of humaning as only you can human. If this is true, the vicar, priest and socialist leader should give up their place at the front, as their example can't have any meaning for us any longer. Instead of these leaders, we might need a psychotherapist or a life coach, someone who holds up a mirror in which we can discover for ourselves what this verb 'humaning' exactly means for us.

What is interesting in this development is the enormous semantic anarchy that it entails for the verb 'humaning'. Imagine how this would work for any other verb (or even for any other word). What would it be like if each of us were to give our own individual meaning to the verbs 'to play football', 'to sing' and 'to clean'. That would be a terrible mess. Incidentally, it would actually be a mess on which we could grow a beautiful Lotus flower together – to anticipate the rest of the argumentation in this book.

The second development is related to the strong scientification of our daily lives. The withdrawal of religiously inspired leaders went hand-in-hand with the rise of scientific professionals. Initially this scientification was limited to the parts that physics, technology and medicine play in our lives, but since the 1960s the social sciences (and since the 1980s the behavioural sciences in particular) have started an impressive advance. These days, every policy measure that entails an intervention in the way society functions needs to have a sound scientific footing. Of course the behavioural scientists take good care that they are neutral, objective and non-normative, as they are conducting science, not instilling morals. Nevertheless, they have had their hands on the verb humaning for a long time, and that certainly has had some morally relevant consequences.

Humaning is not simply an activity that everybody can interpret in his own personal way. It consists of a variety of components that can be studied scientifically, and these studies have produced a great deal of scientific knowledge that we could and should use as a guideline for our humaning. If we want to raise our children in a healthy way, there are many things that we should do and many other things that we should abstain from. Or, to put it differently, and to show unashamedly the moral relevance of behavioural science, if we want to teach good humaning to our children, there is a great deal not only of what they should learn to do, but also of what they should learn to stop doing. It is science that should take us by the hand in this situation, and as a result it is the scientist who has become the proud and confident leader.

This is the development that I am trying to put a stop to with this book – perhaps on my own, as a small David against a gigantic and relentless Goliath. But I may not be on my own. I have pinned my hopes on the philosopher in you, the philosopher in each of us, who has made sure that we adopt an investigative attitude as soon as we find ourselves in a grey area and are at a loss as to how to proceed. In this book I defend the notion that the most appropriate meaning of the verb humaning is exactly this: that we have common sense, that we know when to adopt an investigative attitude. We should not let this ability to be taken away from us, if only because it is one of those abilities that cannot be taken over by other people, which is something that philosophers of the Enlightenment such as Hume, Rousseau and Kant already made clear a long time ago.

You can also say it in the following, simplified manner: we all can and need to 'human' by ourselves, as we are naturally gifted in that, each and every one of us. Moreover, humaning is an activity that is not suitable for what is usually one of humanity's greatest strengths: our ability to use division of labour and do for each other what we as individuals are best at doing. Some people are good at carpentry, and others at gardening, cooking or banking. These differences are wonderful. But just like breathing, humaning is not suitable for a division of labour. We all need to breathe individually, each and every one of us. Just like humaning. Breathing is something that no one else can do for you. Just like humaning. And luckily: breathing is something we can all do ourselves. Just like humaning.

2. Endoxa: a microcosm

In the first part of this book I want to outline the way in which our common sense is grounded in our ability to human, and this involves several skills and attitudes that I will discuss one by one. To illustrate this, I have invented a fictitious small world in which I present a colourful group of people in a variety of scenarios. For this microcosm I have chosen the island of Endoxa. I would like you to imagine that you are living on Endoxa, together with about 100 other people; you can compare it to a kind of holiday camp. These people come from all over, from different, far-away places and from different, far-away eras. In normal life this would of course lead to an enormous confusion of ideas, and for a large part this is exactly what I am interested in: our natural reactions to such confusion. However, I want to introduce this confusion bit by bit, which is why I would like you to imagine that even though the people on the island grew up in all kinds of different places and times, there are no problems with your common daily language. The words that you need throughout the day to explain what you are doing and to understand what other people are doing can be shared in a relatively uncomplicated manner. You could compare it to how things went at your family home when you were about 9 years old and everything just ticked along nicely. That is the situation for you, the people of Endoxa: you all walk around in the same carefree way as you did when you were young. Moreover, in the scenarios that I present you will not be distracted by inevitable cultural, historical and social differences. Together, you have found your feet on Endoxa in a rather uncomplicated manner, even though one of you is a mathematician from the Egypt of the pharaohs, one is an Inuit from seventeenth-century Canada, one is the twentieth-century descendant of the Mapuche in Chile and one is a contemporary reader of English with an interest in philosophy. What you do on Endoxa is humaning, each in his own natural way, without too much cultural baggage. As a 9-year-old child or an adult on a holiday camp, with on the one hand the uncomplicated resoluteness of a 9-year-old at home and on the other hand the open willingness of an adult on holiday.

I realise that this description demands a great deal of imagination, something that philosophers tend to ask when they make use of a thought experiment to support their argumentation. I'll be using life on Endoxa to develop a clear perspective of our daily, natural ability to react adequately to our surroundings. There are three aspects of the interaction between us and our surroundings that I expressly want to take into account:

1. It is clear that we live in an advanced, complex society and that we are usually greatly supported by shared knowledge that is widely available and that is handed to us by experts, often in the shape of implicit or explicit guidelines, such as the manuals for our complicated household appliances. You can see those guidelines as similar to the advice that you can now download from the Internet, for example advice about how to prepare an amazing Christmas dinner, settle an argument with your aunt, replace the battery of your mobile phone, or deal with your senile neighbour.

2. It is clear that it takes years of training to find your feet in daily life. During those years, we learn slowly but surely to make much of our behaviour automatic, and this is of vital importance. During our formative years we start to feel at home in a specific culture, a specific society and a specific era. A great deal becomes normal to us in a very special way: we get used to a particular, individual way of living and reacting.

3. It is clear that we usually live in a stable and predictable environment in which there is a great deal of repetition, of ever-recurring processes. We seem to learn to respond adequately to our surroundings in virtue of these repetitions. We are creatures of habit, and together we make sure that there aren't too many surprises.

In my description of life on Endoxa, I have excluded the first aspect. You are people who live their daily lives on Endoxa without being able to make use of other people's expertise, be it explicitly written manuals or specific advice given by people who simply have more knowledge of life on Endoxa. You will have to live your life on the island without experts. After all you are all strangers, all immigrants, all pioneers. There are no specific scripts for you, and nobody really knows how things work on Endoxa. Instead of resorting to expertise, to fixed procedures, to stable societal structures, you will have to start humaning, to use that verb once again. And what's more, you only have your own skills at your disposal.

Since none of you have any social, cultural or biographical roots on Endoxa, your behaviour will mainly show you that you are able to adapt to new scenarios. This is the ability to recognise a role for yourself in new, unknown circumstances, a role that is suitable in these circumstances and that can be carried out by you. This is the ability that I call common sense. 'New' is the keyword here and, other than you might expect, it doesn't actually have to refer to 'totally and radically new'. It may be radically new on Endoxa, and that may give this thought experiment its added didactic value, but in daily life even a miniscule difference makes a scenario new. The following discussion may help you understand this. At first, every time a baby grabs a rattle, he experiences the rattle as new. He has no idea yet that every time he sees the rattle it is the same one. He cannot differentiate yet between grabbing the same rattle twice and grabbing an identical rattle twice. It is only later that he can make that distinction, when he has become familiar with what Piaget calls 'object permanence'.

It is important to realise that he needs to develop his cognitive skills to see that difference, something that doesn't automatically come along with the actual rattle in his playpen. The baby needs to develop a rudimentary concept of 'rattle' and together with that an underlying, more abstract rudimentary concept of 'object'. Since these two concepts are part of the baby's cognitive skills and are not exclusively attached to the objects that receive their permanence from these cognitive skills, it is sometimes easy to fool children – for example if the pet rabbit has died when you were looking after it. You may well find a new animal that sufficiently resembles the deceased one, and in that way you may extend Bunny's life without being found out. This can be generalised, as a concept always refers to many different objects. As soon as the baby realises what it is to grab his rattle, you can surprise him by giving him your bunch of keys. But after a while he will have adapted his cognitive skills and he will have learnt to deal with a more general, more abstract object. He will have acquired a more abstract concept: 'thing that makes sounds when you shake it'. Even his grandmother's keyring, which he has never seen before and so is new to him, will no longer surprise him: it is a thing similar to his rattle.

The ability to adapt to new scenarios can be found at the basis of the development of a conceptual framework. Conceptual frameworks create order in unknown circumstances by identifying scenarios organised around a role that you can carry out. People who can work with a conceptual framework show that they are able to adapt to new scenarios. The latter is an elementary ability that strangely enough becomes increasingly hidden in our conceptual framework during childhood, as we start recognising unknown objects more easily as familiar things. The fact that you have been socialised, that you have found your feet in your culture and your era, more or less hides your underlying ability to adapt to new scenarios.

I am using the Endoxa thought experiment to distinguish as much as possible between these two abilities. That is why the island's inhabitants come from all over, from different eras, from different societies, from different cultures. That is how I can make a reasonable case for stating that the ability to adapt adequately to new circumstances is not based on the specific way in which we have been socialised, but that it is in fact the other way around: our socialisation is based on this ability. In other words, the fact that we have found our feet in an environment that is familiar to us and that has been adapted to us is a symptom of an elementary, underlying, creative ability.[2] This is the ability to see familiar scenarios in unknown circumstances, the ability to spontaneously match two different things: ourselves and our environment.

There is one specific and characteristic mode in which this common sense manifests itself, and that is the self-evidence with which we adopt an investigative attitude when we are confronted with something that we do not immediately understand. It is this mode that, as a philosopher, I like best, and it is this mode that is most threatened in a society that is increasingly organising itself around the power of expertise.

There are enough examples, especially from the field of advanced technology and its helpdesks. What I mean is this. I can understand the bicycle. I understand what happens when I push the pedals, when I move the handlebars, or what I should do when the saddle is loose. The car, however, is already something completely different. It is much harder to understand what goes on in a car. For example, I'm not completely sure how power steering works. I know it is meant to make it is easier for me to turn the steering wheel, but I have no idea how it works mechanically. What I do understand is that the steering wheel helps me to change the orientation of the front wheels and thus allows me to take a curve. But even this type of simple understanding disappears when we start discussing the Internet or mobile phones. What exactly happens when you surf the Internet? How does wireless Internet work? How is it possible that the computer of the Met office can send its animations to my mobile phone? And how can it do so when I'm on the train and continually changing my position? I don't understand that at all. This makes me lazier, more dependent on the helpdesk. I no longer even try to understand these things, as long as I can understand how to use my mobile phone. And as long as I understand that I can contact the helpdesk if my mobile phone no longer does what I think it should do.

This means that I no longer adopt an investigative attitude in the case of my mobile phone, at least not when there's something wrong with the way it responds. If my mobile phone unexpectedly stops functioning, I will immediately start the missing information subroutine and hand over to the helpdesk, hoping for useful information. Thus, my common sense is shrinking: rather than adapting to a situation in which my mobile phone does not do what I expect it to do, it has become a matter of helplessly putting my fate in someone else's hands. This powerlessness does not necessarily have to be bad. After all, a highly specialised division of labour is necessary in our current complex society with this kind of highly advanced technology.

From here it is only a small step to a cognitive and emotional powerlessness that *is* bad, that really poses a threat to our ability to 'human'. Suppose that our daily dealings have become so complex that we have to conclude that we have no idea what's going on regarding our own behaviour. Suppose we discover that our everyday explanation for our everyday behaviour is completely wrong. Suppose that we have to draw the conclusion that we have no idea at all about the motives behind our own behaviour. Suppose that our ignorance is far greater than even the most radical Freudian ever deemed possible. Suppose that what moves me is so complex and so inaccessible to myself and my loved ones that a gigantic team of scientists is necessary to map my motives. Would that be possible? Is it possible that I'm so complicated, so unfathomable that other people require a helpdesk to understand me and my actions? And if so, will I myself be able to use the helpdesk? Or would I need a second helpdesk for that? And a third, and a fourth, and so on? And could that be possible for all of us? All at the same time?

I hope that these exaggerated questions have given the impression that the suggested scenario is ridiculous. But even if I have succeeded in giving that impression, it is still not immediately clear why this threat of inaccessible expertise is based on a fallacy. After all, some people can be extremely complicated. It is impossible to follow them, for other people and for themselves too. We may be able to imagine that in that sense they are similar to a mobile phone for a non-technical person. And we may also be able to imagine that these experts have insight, on a different, deeper, more correct level, into the motives of anybody else, not just this one puzzling human being but also ourselves, so that it would be better for us to seek their expert advice if we would like to know what is best to do. Which is similar to what we can do if we want to take out a mortgage, draw up a will, file for divorce, fight our depression or want to know how we can best respond to our children. It's the thin end of the wedge, and I think that nobody can predict how far the experts may penetrate into our daily lives and take over our lives with their expertise.

Of course it is hasn't gone that far yet. And it is never going to go that far. At least that is what I'm going to make a case for in the following chapters. But there is reason for concern, as the more we get used to the presence of experts in all corners of our existence, the more we will get used to the expectation, when we are confronted with even the smallest problem, that other people's expertise will bring a solution. The more we admit expertise into our daily lives, the more we will start the missing information subroutine and the less we will tend to adopt our investigative attitude. This means that our lives would become unimaginably poorer, and there is no need for that at all.

You will notice on Endoxa that you can get far using only your common sense, without any expertise. You will notice how refined the natural casualness is with which you can adapt to new scenarios. You will notice how multi-faceted our daily common sense is, how much is involved, and how much power it has. You will be surprised about all the things you do and can do without expertise. You can only see this if you focus your attention on it, if you aim the spotlight at our common sense, and this is something you cannot escape on Endoxa. You will be wondering about your natural, everyday talent, just like you may be amazed at all the things you can do with your hand − the type of amazement you only feel in special circumstances: the day your RSI has been cured, your pot is removed from your broken wrist, or your numb hand slowly comes back to life.

Notes

1   Cf. Aristotle, *Ethica Nicomachea*, 1178b, 7.
2   This is a major theme in Ernst Cassirer's philosophy of symbolic forms. See Cassirer (1923–9). Some echoes of this can be read in Goodman (1978).

2 Expectation between prediction and hope

1. A ferry tale

Suddenly you wake up in the middle of the night. There is somebody next to your bed. It is all a bit strange and indistinct. Even though you know it is your bed, it doesn't look like your bed and it doesn't look like your bedroom either. Why is that woman there? She starts speaking to you and suddenly you realise that you know what she's going to say.

— Come on. Wake up. I've come to pick you up for the experiment, remember? The boat leaves in 20 minutes. Come on.

A little bit later you're walking through the cold night. It isn't completely dark anymore; you're walking in the early twilight. In the bluish light you notice that there are more of you. It seems as if someone is joining you at every street corner.

Then there's the port. The water laps against the quay and there isn't a ship in sight. Oh, but there's one right beneath you. The quay is much too high for the small ferry that you can make out beneath you. The vessel is swaying in the waves and keeps knocking against the quay. Surely that little boat is much too small for the crossing. You're shivering and all of a sudden you feel lonely and lost, a long way away from home.

You turn to face Brenda and yes, you realise you should have known: the woman who has just woken you up now resembles Lupus, the werewolf from Harry Potter. It is his ship. You're not surprised, but you are beginning to feel worried.

And the boat is still moving up and down quite violently.

You are all standing tightly packed while the little ferry slowly leaves the harbour, sailing into a dense fog. You can't see a thing. You can feel how your hair is getting wet, how it is sticking to your forehead, how the drops are forming. One of the drops is slowly sliding down your nose; you wipe it away with the back of your hand.

The boat is swaying heavily. You need to hold on to the railing. Someone has grabbed hold of your arm. In front of you there is a huge man with a loud shirt on, facing away from you.

That's funny. You realise that so far you haven't paid any attention to your own clothes. When you do, you are rather embarrassed as you are wearing a similar Hawaiian shirt. Where on earth did that come from?

Everybody around you is wearing similar shirts, you now notice. What a cheerful group of people!

Then suddenly the fog lifts, and it is gone just as soon as it appeared. People start cheering and you see, right in front of you, the most fantastic tropical beach that you could ever imagine. Wonderful! Now that the fog has lifted, you can feel the pleasant warmth of the sun. What beautiful light! You're going to have a wonderful time. You have arrived!

— Welcome! Welcome to Endoxa.

This is Esteban, warmly welcoming you. He's calling to you from the jetty. He reaches out to you with his hand, and while the small boat is moving up and down, you grab his hand and step onto the jetty.

— Hello, you say. Finally.

You smile broadly. It is about to begin.

2. Folk physics

Imagine that you have been invited to stay on Endoxa. Imagine that you have enough adventurous spirit in you to really look forward to that trip. Imagine that you've heard some of the details, but that you're not exactly sure what it is all about. They have told you that you're going to a sunny island, with all kinds of different people unknown to you: people from various cultures, societies and eras. This must feel odd, very odd. It must be nearly impossible to start imagining a situation like this. The different eras are especially difficult to take in. After all, there are so many practical hindrances. But then again, in today's quasi-virtual world of games, computer-animated films and 3-D projections, many things that may seem inconceivable are actually quite possible. So let's suppose you say 'yes' to the trip. You'd probably think it would be something like a holiday colony, like in reality TV programmes such as Big Brother, Fort Boyard and Expedition Robinson. You have some ideas about what it could be like, but none of these are very concrete. You'll just wait and see what happens.

Your trip to Endoxa was full of strange occurrences, things that cannot be real. After all, your bed never finds itself in a strange room all of a sudden. Anyway, how can it be your bed if it doesn't even look like your bed and it is not in your room? And people do not simply change into Harry Potter characters, nor do you simply know the names of people you have never met before. It feels strange. It feels very much like a dream.

In addition to these strange, bizarre and impossible occurrences, there are also many completely natural things that happen. You can hear the lapping of the waves against the quay wall. You can see the little boat bobbing on the water. You can feel it rocking, and you need to hold on tight. Someone else on board is holding on to your arm. That person must be somewhere in the middle of the vessel, in a place where there's nothing else to hold on to. Don't you think? You must have thought the same. After all, strangers don't usually hold on to each other's arms. It is likely that you have added this information to my description without even thinking about it.

Why is everybody packed up so tightly? Are we all afraid of the werewolf at the helm? Is everybody trying to keep as far away as possible from it? Or is it simply because there's no more room? The people in the boat may have instinctively divided up the available space into equal parts, like you do in a lift. If there is only a little space for many people, you automatically stand closer together. Is there is a lot of space for few people, you automatically stand further apart. That is the way we do things. Equal division is natural. It is comparable to pouring water into a bowl: the water spreads over the surface rather than pooling in a corner. This is the way water behaves. And the same is true for you as a group on that boat. It has little to do with psychology; it is simply a question of how human beings behave in certain spaces: similar to water molecules in a bowl, starlings in a flock and cars in a traffic jam.

The knowledge that this is just the way things go, that cars, starlings, molecules and bodies behave in such a way, is an extraordinarily interesting kind of knowledge. It is an example of *folk physics*: the knowledge that people have gained in their daily lives about the behaviour of the things around them. This knowledge is displayed in the expectations we have about the things that we have learned to deal with. For example, think of the fog around the boat. Notice how you wiped the drop of water off your nose, without even giving it a thought. Notice how you stood firm and held on to the railing. This is knowledge that is embedded in your muscles and your spine, rather than in your eyes or your mind. Your body knows how to keep its balance on a boat that is rocking on the waves.

It is important to realise how amazingly clever this really is. Notice how you step from the boat onto the jetty and how smoothly your hand moves forward, exactly far enough, how you take hold of Esteban's hand, firmly but without squeezing it too hard, how you move your weight from one leg to the other, how you pull yourself up, step onto the jetty and then let go of Esteban's hand. And notice how the two of you do this together, like a graceful dance. How Esteban puts out his hand, grabs your hand, braces himself, and moves his weight too, without falling over. And all this without even thinking about it; it is really phenomenal. If it weren't such a commonplace occurrence, it would merit a standing ovation.

This mindless skill demonstrates your own folk physics, just like your knowledge of fog, of your wet hair sticking to your forehead and forming a droplet, which then slides down your face. A drop like that will never slide upwards and move through your hair all the way up to the crown of your head. You have probably never even considered that. But of course you know how drops of water move downwards and how the heat of the sun feels after the fog has lifted. You deal with this heat as a matter of course, just like you do with the sound of the water splashing against the quay and the fog impeding your vision.

Gaining experience with the cause-and-effect relationships between things, each and every one of us has learned to find his way as an embodied creature. This learning starts early, when we are stacking building blocks on top of each other, before we have even learned how to stand up. For example, we learn that a ball does not lie still on a slope. We discover that a rolling ball does not suddenly go round the corner. And if somebody spills soup on the table near us, we quickly move backwards so that our clothes don't get dirty because we know that the soup will not simply stay there but will flow over the edge of the table and may then soil our clothes. But if the table has a raised edge, we will stay where we are. We do not need to think about it, that is just the way it is. Soup cannot run up a raised edge. That is not what soup is like, and we know this. In an exceedingly sophisticated way, this is part of our body's knowledge. Just like we know that we cannot look around the corner, but we can hear what's happening around the corner. And so on. As a matter of course, human beings have an incredible knowledge of the behaviour of the things around them.

Not only do we have knowledge of the things around us, but also of the people around us. Throughout the years our bodies not only assimilate folk physics, but also folk psychology. Esteban reaches out to you and you react automatically. He wants to help you ashore, and you understand this. Somebody grabs hold of your arm, and you allow this because you understand that if you don't, this person will fall over and most likely he doesn't have anything else to hold on to. Just like that huge man in front of you, you are wearing a loud printed shirt. Granted, you may not feel completely at ease in this shirt, but what if I had left you naked? You would've been quite surprised if I had made the off-hand remark at the end of the story that you badly scratched your bare buttocks when you left the boat because you didn't see that rusty nail sticking out. Even if I hadn't mentioned clothes, you would have assumed you were dressed. That's the way people do things, and you just know this. And just as we are becoming familiar with the folk physics of our world, we are becoming more familiar with the folk psychology of our world. Our folk physics and folk psychology help us in our 'humaning', to use that verb once again.

3. Expectations, entitlements and obligations

There are not only interesting similarities, but also interesting differences between folk physics and folk psychology. The first similarity is that both forms of knowledge are part of people's expectations. Human beings have been part of the passing of time, of the unfolding of scenarios. Human beings know what to expect of the things and the people they will be dealing with. Both our folk physics and our folk psychology are realistically building our expectations, anticipations and habits. They constitute our knowledge of experience, knowledge of the way scenarios unfold and knowledge that is firmly and immediately embodied, for example in the way our body braces itself on a rocking boat. This is the knowledge that allows us to walk, because our weight is automatically shifted from one leg to the other, that allows us to shake another person's hand, that allows us to grab a railing, and that allows us to pass other people in a narrow corridor. This knowledge even allows us to walk up the stairs carrying a big box without being able to see the steps, and then we can even switch on the light, unthinkingly, with our nose, because we know where the light switch is. Move house and you will see. And if you go and move to Endoxa, you will notice how much implicit, local knowledge moves along with you in your everyday expectations; you will find that this is knowledge that only slowly ebbs away when your new dwelling place requires different expectations.

There is a second similarity between folk physics and folk psychology, a rather subtle similarity. Both types of implicit knowledge are based on experience, of a history going back years, although it is not important whether this is an identical, specific, shared history. What I mean is the following (and this is easier to explain by means of folk physics than by means of folk psychology): it is possible that I learn the behaviour of things through my experience with for example bananas, phonebooks and key rings, whereas you learn the behaviour of things through your experience with for example oranges, beer bottles and thimbles. There are overall cause-and-effect patterns that are endlessly repeated in your contact with all kinds of things, such that even a relatively limited and exotic collection of things can still help you become acquainted with these patterns. For example, things take up space. In the place where one thing is, there can't be another thing. And it doesn't matter whether you live in Santiago, Samoa, Stockholm or Southampton, whether you were born just after the Second World War, in 1983 or in the year 1043, you will develop enough experience with physical objects to develop roughly the same folk physics.

And it is the same with folk psychology. I may learn to 'human' through my contacts with for example Dave, Catriona and Harriet, whereas you learn to 'human' through your contacts with Gustavo, Amanda and Esteban. There are motivational patterns that are endlessly repeated in the way we 'human', the way we react to each other, so that even a relatively limited and exotic collection of people can still help you become acquainted with these motivational patterns. Human beings want to be understood by other people and by themselves; they want other people to be able to follow what they are saying, and to agree with what they are doing. Wherever and whenever you are living, you will always have enough experience with people to develop roughly the same folk psychology. Therefore, even though you share no specific history or specific set of instructions with the other people on the island of Endoxa, you will still have sufficient experience of human beings and the way they behave to have roughly the same folk psychology as your fellow island dwellers.[1]

I must, however, quickly emphasise at this point that there is a fundamental difference between folk physics and folk psychology. This difference is closely related to the difference between the knowledge of cause and effect and the knowledge of rules and reasons, and therefore with the various obligations and entitlements that are related to our expectations. Please allow me some time to explain this using two simple examples.

Stone

Lifting that stone made an enormous impression on you. It was a big, grey stone and it was in your way, partly blocking the door to your cottage. It had to go, so you braced yourself to lift it. But what a strange surprise, what an odd sensation! The stone was in fact pumice stone. You almost fell over backwards because it was so light. Sumalee saw you do it, and after she had given up on the idea that you were incredibly strong, she started playing with the stone herself, as if it was a ball. She soon discovered that it could even float on water. She just couldn't get enough of it: amazing that something like this could exist, a stone floating on the sea!

Date

It is quarter past six. Dick has been waiting for half an hour in the shade of the big palm tree near the turtle beach. He's got a date with Gabriella. He likes her so much; he thinks she's so sweet. The way she smiles! Dick feels wonderful. He is still glowing a little bit, proud of having asked

Gabriella to meet him. She said she wanted to meet. She said she would be there. And then she had smiled that smile again: so gentle, so charming. Dick had put on his best T-shirt and now he is sitting there under the palm tree, slightly restless, slightly nervous. He knows that he got there way too early, but by now he is certain that Gabriella is late. He feels a slight panic building. And then he sees her coming. Ah, good!

But what is this? She's not alone. Gabriella is coming his way, but Jayden is walking alongside her. Jayden is coming with her. What on earth is he doing here? What is going on? Didn't Gabriella understand?

The expectations in the scenario *Stone* are based on folk physics. You know about cause and effect and you know how heavy stones are. You know how much energy is needed to lift such a big stone. But in this case, it turns out to be no ordinary stone: this stone is very easy to lift. Sumalee's expectations are also based on folk physics. She thinks that you must be really, really strong. And then it is fascinating and fun to throw the big stone into the water and to discover again and again that it just stays afloat.

By contrast, the expectations in *Date* are based on Dick's folk psychology. He has a romantic date with Gabriella. But when she shows up, she's not alone. Apparently she hasn't understood, and neither has Jayden. Or has this man got his eye on Gabriella too? Is Jayden trying to outdo Dick?

Expectations based on folk psychology do not demonstrate knowledge of cause and effect, at least not primarily. They mainly deal with knowledge of human motivation, of knowledge of the rules that human beings tend to follow, either because they want to or because they feel obliged to follow these rules. That is why these expectations also involve people's emotions and their motivation to do the things they do. These psychological matters create a type of relationship that is completely different from a cause-and-effect relationship. This is why there are two things that occur when people start 'humaning', when human beings meet each other as human beings, when they rely on expectations based on psychology. On the one hand, everyone realises that this type of scenario does not deal with strict patterns, but rather with rules that people *usually* follow. On the other hand, everybody realises that the expectations in question have a normative value. People are supposed to adhere to this kind of expectation and this is something that you should be able to count on. Humaning, the verb, is a question of adhering to the correct obligations and entitlements.

The terms obligations and entitlements may sound rather formal and business-like, but they are still the right words to use.[2] Try and imagine how you would feel if you were romantically waiting for Gabriella under that palm tree, and you saw her arrive with Jayden. Would you be disappointed? Angry? Would you be embarrassed about your naivety? Or about your romantic feelings? None of this would be surprising. How you would actually feel depends on the obligations and entitlements that are embedded in your expectations. If you are angry or hurt, it is because you feel that you are entitled to expect a romantic reaction from Gabriella. And this is the situation for Dick. He thinks that their date means that Gabriella has taken the obligation upon herself to play her role in a romantic scenario. However, Dick might also feel embarrassed, realising that he had indeed assumed (wrongly?) that he was entitled to his expectation that Gabriella was also expecting a romantic scenario. Now that it appears that they don't have the same expectations, Dick realises that there were insufficient grounds for him to think he had this entitlement. This is why he is embarrassed: he suddenly realises that Gabriella has an entitlement of her own, namely the entitlement to consider the date as a neutral, strictly friendly meeting. And this entitlement obliges Dick to take it slowly.

In other words, in *Date* there are no fixed causal relationships, just complex and subtle interactions between people who are trying to interpret the situation they are in. Dick's expectations are based on his assumptions regarding both his and Gabriella's entitlements and obligations. These assumptions demonstrate his interpretation of the situation as a romantic scenario. However, Gabriella appears to have had different expectations, implying different entitlements and obligations. Her expectations do not match Dick's, and vice versa. Their expectations clash and it is not immediately clear who will need to adapt his or her expectations, who can lay a claim to what entitlements, and who needs to assume what obligations. For Dick, the lesson learned on Endoxa is that he needs to be more careful with his folk psychology. Entitlements and obligations that are obvious to him may not necessarily be valid on the island of Endoxa. On this island, humaning is really something you need to work on: it doesn't happen of its own accord, and it's not something you can do on autopilot.

In *Stone*, you learned a lesson regarding folk physics. Here too something unexpected happens, because that stone in front of your cottage turns out to be pumice stone, which is not heavy but light, even to such an extent that it floats on water. But in *Stone* it is not a matter of ill-adjusted entitlements and obligations. The stone has no expectations of you, and your expectations of the stone are not normative. They may indeed deal with what you consider normal behaviour for a stone, but this normalcy has nothing to do with what it is *appropriate* for the stone to do. What you expect of this stone, what you think is normal for stones, is only a matter of your knowledge of the strict behavioural patterns of stones.

The fact that the behaviour of pumice stone doesn't meet your expectations is no reason to think that there are no causal behavioural patterns at all for this stone. Of course there are. The behaviour of this piece of pumice stone doesn't affect your folk physics. You simply need to realise that the stone is much lighter than it looks, at least than it looks to someone like you who has learned the behaviour of stones in a world without pumice stone. Once you realise this, there is no problem. The stone then resembles a plastic ball, and this is something that Sumalee immediately seems to pick up on.

Incidentally, the fact that Gabriela doesn't meet Dick's expectations cannot be a reason for Dick to assume that Gabriella doesn't respect *any* entitlements or obligations at all. Of course she does. How else could she do her humaning? Of course Gabriella follows rules, rules that usually apply. In that sense, Gabriella's unexpected behaviour does not affect Dick's folk psychology, at least not in its fundamental core, which is the understanding that you need normative regularities in order to be able to do your humaning. There is one thing that makes our folk psychology different from our folk physics, and this is something that makes humaning really into an activity that you can do well or not do well. This is the fact that folk psychology is much more flexible, and that it is the result of us successfully creating a common world together.[3] On Endoxa this is not just a simple fact. On Endoxa humaning is a verb, an activity.

4. Life without experts

By curious coincidence you are now spending your days humaning on Endoxa, together with one hundred other randomly selected people. You are all the product of my thought experiment. I am using you to gain insight into how people do their humaning, without experts but using common sense. All you share on Endoxa is your folk physics and your folk psychology. You all have a global idea of, on the one hand, cause-and-effect relationships between things, and on the other hand normative relationships between people. You all have particular expectations of what you can expect in certain scenarios. You all have a rough realisation that one occurrence leads to another and that there are certain natural laws that control this. Moreover, you all realise that when you and your fellow visitors react in a certain way, if it is to do with humaning, then it is not simply the laws of nature that are at play. In those cases it is mostly a matter of habit, of the things you consider normal, and of obligations and entitlements: what people usually do and what they can expect of each other.

This realisation is indeed quite a step already, but since everyone has their own past, their own ideas about what is and isn't normal, and has experience only with their own limited and exotic collection of things and people, there is still a great deal to be explored and discovered. If you are Canadian, you will be blown away by the sudden nightfall on our tropical island, as well as by the heat, the lush green plants, and other natural characteristics. These will not surprise you at all if you hale from the Egypt of the time of the pharaohs, but then you will be surprised by the extraordinary folk physics of your twenty-first century fellow visitors. And you may also be equally surprised about their weird folk psychology. You'll have to get used to their reactions, you'll want to find ways of expanding your own folk physics, and it will be in your best interests to expand your own folk psychology. But how are you going to approach this?

For me it is interesting first of all whether you will realise that there is such a great difference between the expectations based on your folk physics and the expectations based on your folk psychology. This may seem like a natural difference, but is it? Let's have a look at the following scenario.

Tears

It's been dry on Endoxa for weeks, and stiflingly hot. If you can sleep at all, it is only briefly. There is bound to be a water shortage. The grass has turned yellow and brown, and the goat has become even bonier than it was and is constantly bleating plaintively. Anchor, a man you like very much, hears the goat too and then gives a wonderful imitation of the sound. The animal's sadness echoes so deeply in Anchor's voice that it touches you deeply. You can feel the tears rising, and then you give in to them. You start crying, weeping with emotion, with sadness, with pity. Warm tears stream down your cheeks. How good this feels, and how confusing. Ningak sees you crying. He understands: it's the heat, the exhaustion, the goat bleating, Anchor's imitation. One thing calls forth another. Rebecca can also hear you. What a drama queen, she thinks. Do you want attention so much that you need to attract it in this way?

Rebecca thinks you have the choice to cry or not. She holds you responsible for your crying. She thinks it is not normal, at least not normal for people like her. It may be normal for people like you, cry-baby. For her, it is simply a matter of folk psychology: you wouldn't expect such visible emotion from normal people. You ought to be ashamed of yourself!

To Ningak, the situation is completely different. Just as the goat can't help bleating, you can't help crying. It is only to be expected. It is simply a matter of folk physics, or rather of folk biology. If you are so exhausted, if it is so hot, Anchor's striking imitation of the plaintive bleating is simply the last emotional straw. Then you *need* to start crying; it is simply a matter of course.

And you? You don't know. You like Anchor a great deal, and you realise that this is also part of it. He touches you emotionally and you did feel the tears coming. Perhaps you could have stopped them if you'd wanted to. But does that mean that you deliberately started crying? Or were you yourself surprised by your emotional outburst? Did you allow it to happen, did you make it happen? Or did you have no choice in the matter and did it simply have to happen? It seems like it is both one *and* the other. Confusing.

The distinction that I make here between folk physics and folk psychology is a theoretical distinction. It is a Western one, the product of modern science. It is a distinction made from an impartial, detached perspective, and perhaps this perspective is necessary to make it convincing. This is a distinction that is made intuitively in everyday life, in a comparable but slightly different manner, between meaningful and meaningless processes, between mechanisms and people, between expectations based on the laws of causality and expectations based on sound reasoning.[4] All the ways to make a distinction between the two different types of expectation (to use my favourite characterisation) are partly evident and lucid, and partly problematic and murky.

Clearly, things can be either black or white. The pumice stone is not out to fool you, it is not out to do anything, nor is that what you expect. In the scenario *Stone*, all your expectations are causal. In the scenario *Date* that is clearly different. Dick has an ulterior motive; he has normative expectations. You can see this in the way his expectations are emotionally charged. That is clear: black or white. But a scenario can also be confusing, murky, fuzzy and grey, like the scenario *Tears*. Did you start crying on purpose when Anchor gave his striking imitation, did you deliberately let your emotions take over, or did you have no control over this and did you simply become awash with emotion? Is Ningak's causal interpretation of your behaviour correct? Or is Rebecca's normative interpretation correct? Or are neither? Or perhaps both? So how should we interpret the situation then?

I'm not sure. Life is often murky, confusing; the world is often multi-interpretable, astounding, overwhelming. There are many situations similar to the scenario *Tears*. And in a way there are some similarities between *Date* and *Tears*. A cultural anthropologist may well easily explain to us how things simply *had* to go wrong between Dick and Gabriella, that in *Date* a causal interpretation is more appropriate than a normative.

In other words, there are a lot of grey areas and these grab our attention. These grey areas should amaze us, as they merit an investigative attitude from us and from the people on Endoxa. Who knows what will happen? Rebecca may be able to expand the domain of normative expectations more than seems possible. Or more than is appropriate. And Ningak may reduce the domain of normative expectations more than is appropriate. Or more than seems possible.

This is something you and your fellow visitors will have to decide together. You will need an investigative attitude for this, as this is all you have on Endoxa. The missing information subroutine is no use to you on Endoxa, as there are no experts who can simply tell you. You will have to do your own humaning. Together you will have to make your own common world liveable.

5. Looking for support: predictions

In our current highly developed knowledge society, experts have known for a long time that our folk physics needs to be replaced by modern natural sciences. Folk physics may be alright in daily life, when we are mostly dealing with medium-sized objects and gravity. But the experts know that folk physics is limited, as it cannot deal with all kinds of complicated and subtle physical processes, and it is often downright wrong. That is why it is a good thing that we have science. That's why it is a good thing that there are experts, people who really know how the underlying mechanisms work, the mechanisms that we try to deal with using our folk physics that is based on our intuition and experience.

I am not going to argue against this. Without modern science our life would be much more primitive and, from a material point of view, much more deprived, difficult, painful and sober. And it is not only high-tech stuff that we have science to thank for: even simple basic things like a sewage system and clean drinking water are the results of science. In that respect I am slightly worried about you people on Endoxa. After all, I wish you too could have a smartphone, an economical heater, Internet access, an e-reader, as well as a hypermodern hospital in which they can perform great endoscopic surgery. Unfortunately there are no such things on Endoxa. However, I was kind enough to make sure there is sufficient hygiene, healthy food, warm clothing and pleasant accommodation (or I would not have obtained approval from the Ethics committee!). In fact, I am not so much interested in your folk physics but rather in your folk psychology, in the role that experts play in your daily communication. I am using you on Endoxa to clarify the kind of expertise that could be extrapolated from your folk psychology.

If we extrapolate from your folk physics, we end up with the modern natural sciences. Of course there are great differences: modern physicists are concerned with totally different things than those we use in our daily lives to build our experience. The things that physicists study are minute, nano-sized; you can't really call them 'things' anymore. They are waves, oscillations, snares, charges: I'm not really sure what exactly. But even though these 'things' are incredibly small, even though they are no longer real things, what has remained is that it is still all about a series of events. Even though current ideas about cause-and-effect relations are incredibly complex and not even linear anymore, in essence it is still about the knowledge to correctly predict what will happen. In that sense, the modern natural sciences are an extrapolation of our folk physics. And it is because of the formidable predictive power of natural scientific theories that we have the tendency to replace our folk physics by the more exact and successful natural sciences.

The relationship between folk physics and modern science can be explained further by emphasising some of the connotations of the concepts EXPECTATION and PREDICTION. Expectation sounds subjective: it refers to mental states that are evoked in us by the way things usually manifest themselves. It has no theoretical foundation, but it has grown in practice and in our experience. Prediction on the other hand has the pretension of objectivity. It is not based on practice or daily experience, but on privileged access to knowledge. In popular wisdom prediction is associated with fortune tellers, double-dealing and cheating, and it is this association that gives prediction its special reputation. There are two sides to this reputation: prediction demonstrates an extraordinary ability to look beyond the immediately visible, and it is exceptionally valuable because it provides certainty about future events. This exceptional ability is based on swindling when it comes to fortune tellers, but in science it is based on methodologically responsible and theoretically underpinned observations. Natural scientists know what they are talking about: they can make genuinely reliable predictions.

We can also rephrase it as follows: predictions allow us to build firmly on an objective foundation provided by fact and theory, whereas expectations force us to appeal strongly to our own interpretation of ambiguous information. Both predictions and expectations provide us with the opportunity to anticipate future events; however, expectations provide us with less certainty than predictions. In science, this contrast can nowadays also be found if the concept EXPECTATION is associated with statistical probability. If things behave fundamentally in a chaotic manner, for example in the atmosphere, we can make less far-reaching statements and need to make do with expectations. Prediction is not one of the possibilities in the knowledge domains that primarily deal with probability.

These observations regarding the concepts of EXPECTATION and PREDICTION can be used on the one hand to underpin the premise that the modern natural sciences are an extension of our folk physics, and on the other hand to underpin the premise that we would like to replace our folk physics by the modern natural sciences.[5] After all, the latter do the same things as our folk physics, only better. The natural sciences have liberated us from the doubtful interpretations that we sometimes had to use in our folk physics, and as such, the natural sciences have strongly improved our view of physical reality. They have offered us the possibility of replacing our subjective expectations with objective predictions.

Of course, the following question now presents itself: what about the domain of human interaction, in other words the domain of folk psychology? Can science play a role here, too? Can the behavioural sciences, as an extension of our folk psychology, help us replace our subjective expectations with objective predictions? And is this what we want? Are the normative expectations that are characteristic of our folk psychology in fact incorrect predictions based on a subjective interpretation of ambiguous information?

6. Hope: hearts beating in anticipation

There is one meaning of the concept of EXPECTATION that I haven't dealt with so far. That is when expectation has something hopeful, something we encounter for example in Christmas songs such as 'Santa Claus is coming to town'. Of course we know that Santa is coming. We are absolutely 100 percent sure. And of course he has also brought presents for us. Of course! We are absolutely 100 percent sure. But still . . . But still.

This is not a prediction. Is it a kind of certainty based on the laws of causality? No, of course not. That is not how it works with Santa Claus; that is the whole idea. Santa Claus is kind. Santa Claus is good. That is why we are simply certain that he will come. And that is why we are equally certain that he has brought presents for us. But exactly for those reasons, because Santa is good and kind, and our expectation is based on that, that is exactly why we are so full of excitement. Because it is a matter of his goodness. We need to surrender to that: we are placing our fate in his hands, giving up any control we might have. No predictions, no laws of causality, but trust, complete certainty, surrender.

This is an instructive pattern in which our hope is related to our expectations about other people's behaviour, expectations based on our folk psychology. We are sitting by the fireside singing beautiful Christmas songs. We are full of confidence and fervently hoping that this scenario will unfold in the way we so deeply desire. This expectation, this hope, this trust is not based on scientifically underpinned predictions or on knowledge of the laws of causality. That would ruin the whole thing. Such knowledge would rob the scenario of its heart and its meaning, as one part of our deepest desire is that Santa Claus has decided to bring us presents *completely of his own accord*. The presents and appreciation need to come from Santa Claus himself and should not be the consequence of any law of causality. Santa Claus is not a machine in which you insert some songs to cash in on your presents.

Now let's compare this Santa Claus scenario with *Date*. Dick seriously expects Gabriella to show up on her own. He is romantically waiting under the palm tree and assumes that Gabriella will understand what is expected of her. He counts on Gabriella's ability to independently, sincerely and correctly deal with the obligations and entitlements involved in a romantic date. Especially because he hopes to receive Gabriella's sincere and authentic confirmation of his expectations, Dick's heart is beating in anticipation. That is all part of it. If he can't deal with that, if he wants more certainty, if he likes guaranteed predictions, if he wants to control social interactions with his knowledge of the laws of causality, he had better stop dating. Perhaps one of these life-size blow-up dolls would then be better for him.

Let's now go back to *Tears* for a moment, that fuzzy, murky, confusing scenario. What about you crying? What about Ningak's thoughts? Was it indeed simply a biological reaction? And what about Rebecca's thoughts? Was it simply a matter of attention seeking?

In *Tears* these are exactly the questions that merit your attention. These questions revolve around understanding, questions for which you need an investigative attitude rather than a missing information subroutine. They are questions from which you cannot run away, that you cannot send on to experts, exactly because these are questions about *your own interpretation* of the scenario.[6] This is the scope of the expectations based on your folk psychology. Such expectations are not half-baked predictions that we need to use from a lack of certainty, but rather openings towards your engagement, towards your involvement, towards your view of the scenario. *Tears* is a scenario in which people deal with other people and try to build on the other person's folk psychology. That is what such scenarios are all about. Both *Tears* and *Date* are about building a common, meaningful world based on each other's folk psychology. What is needed for this is the subject of the next chapter.

Notes

1   This is, basically, a pragmatist theme. Cf. Dewey (1922) See also Baker (1999).
2   I have adapted these notions from Brandom (2000), who uses them to explain the social nature of linguistic competence.
3   Adam Morton brilliantly explores the relation between the demands of human cooperation and the nature of folk psychology in Morton (2003).
4   The distinction and its naturalness in our contemporary thinking about how we deal with the world in our thought and action is the main theme of McDowell (1994). My presentation of this distinction in terms of distinct kinds of *expectations* echoes my pragmatism, congenial to Dennett (1987) even though I consider it to be congenial too to McDowell's claim that the distinction is to be understood in terms of human spontaneity, i.e. human orientation towards the world.
5   The idea that science can *replace* common sense is based on the mistaken idea that common sense is some kind of 'proto-theory'. See Baker (1999). 'Replacement' is here suggested to be rather radical, as if we are talking about incommensurable paradigms. Cf. Kuhn (1962).
6   I take this to be the major theme of Dennett's work over the years, ever since Dennett (1987), and even before. Even though he is often interpreted as a reductive instrumentalist, he is much more of a serious Socratic kind of philosopher: scenarios such as *Tears* deserve an investigative attitude, and no greedy reductive attempts, neither to a scientistic physicalism nor to a chauvinistic dualism.

3   Understandable, efficient and good behaviour

1. Everyday scenes

She looks at me, and then she says:

— If you like, we could also go left to the turtle beach, past the Birdman. I don't mind.

— Oh no, that is not what I meant. I was just looking at that path. The way the sunlight falls on those trees is so beautiful that it caught my attention. Did you think I . . .

— Well, I wasn't sure myself, I think. We could go straight ahead here, past the campfire, but then again, we could also go left past the Birdman. Then you pass by the Old Fort, the ruins. But straight ahead is shorter, and there is more shade that way.

She is silent for a moment.

— It just occurred to me: 'shorter'. Why on earth would we take the shortest route? We are not in a hurry, are we? I was going to show you the entire island. And when I saw you looking at that path towards the Birdman, I wasn't sure which way to take. That's why.

The last utterance follows hesitantly. The way Sybille talks is endearing. So much detail.

The heat is everywhere; I wipe the moist off my brow.

I am a guest on Endoxa just for one day. It is an odd experience, as if I am an artist who has met himself in his own drawing. Sybille knows that the idea for this experiment comes from me. She wants to show me the island.

We go left after all. The surroundings are fantastic, overwhelming. All those colours, those exuberant flowers and birds everywhere. The Old Fort is completely overgrown. Lizards scoot away on our approach. A man is standing in front of us, bent over. What is it that he has got there? It is a big dead bird lying at his feet, cut open. He leans over really far, as if he is smelling the bird's innards. He has probably heard us approaching, as he suddenly straightens up and looks at us.

— Oh. Erm, I was wondering how it could have died. I found it just like that. An animal? Poison? I'm not sure yet what it could have been.

— Hi Jayden. This is Jan Bransen, you know, from the experiment.

— Ah yes. Of course! How good of you to come and take a look for yourself. Have you come to visit the Old Fort? I'd be happy to show you around.

— Well, actually we are on our way to the turtle beach. Anthony is there and has something to eat for us. Fish, I think.

Sybille and I continue on our way. It is not a big island, about as small as Jersey. You could easily walk around it, at least if it wasn't so hot. We pass the Birdman, which turns out to be a 10-foot high rock, in the middle of a field. It is an unusual sight, a rock which looks like a tall man with a bird on his shoulder. Close by a goat is bleating plaintively.

Someone out of sight is imitating the goat, and it sounds surprisingly natural. Sad, sensitive, with a delightful full sound. A young woman sitting near us on a stone swallows deeply and then bursts into tears, clearly touched by the sound. I notice that her cries move me too. All of a sudden the whole situation is extremely emotionally charged.

— Ah, what's the matter Dituju? Why don't you come with us? We were just going to have something to eat at Anthony's, on the turtle beach.

Sybille puts her arm round Dituju's shoulder and takes her along with us. The sea turns out to be very close. A wide beach, so bright that the sand actually looks white. Under a parasol a man is bent over forward, looking inside a cool box. Funny. The way he is standing is similar to the way Jayden was 10 minutes earlier, when he was inspecting the bird's innards. The man is frowning, holding the lid of the cool box in his left hand.

— Hi, Anthony. What's the matter? Have you forgotten what you need?

Sybille is a woman who notices things. Anthony looks at her and I can see in his face that she is right.

— Oh well, that happens to me all the time, I say. My name is Jan Bransen. Pleased to meet you.

At first Anthony looks at me sheepishly, but then his face brightens up into a broad smile.

2. Rationalising

People are talking animals, *zoon logikon*; this is something that the ancient Greeks already knew. People comment on their own behaviour, illustrate it, put it in context, and explain the intentions of what they do. It is often said that people rationalise their behaviour. They interpret their behaviour, such that it seems to make sense in virtue of their underlying reasons.

Behaviour is supposed to be understandable, efficient and good, as I will explain and defend in this chapter.

Have a look at what Anthony is doing. Bent over, he is looking inside the cool box. I can see him doing it and immediately I associate this situation with a particular scenario. I suppose that he is preparing food and needs an ingredient that has to be kept cool. He can't find it straight away, and that is why he is having a good look inside the cool box. Jayden's behaviour, on the other hand, is less understandable. His posture was similar, but it is unlikely that he needs something from inside the bird. A bird is not something that you preserve things in. Of course, there is a possible, fitting scenario: Jayden found a diamond and thought that the best place to keep it would be in the intestines of a dead bird. Unlikely, but possible. Still, this is not the first scenario that comes to mind. I recall Evans-Pritchard's study of the Azande, an ethnic group in North Central Africa that inspect the innards of birds to discover the cause of unexpected illness in their tribe. Could Jayden be a Zande? I have no idea.

Fortunately Jayden is faster than I am, and that is normal too. Jayden likes to rationalise his behaviour, as we all do. Jayden would not like us to get the wrong idea about him, which is why he immediately explains his behaviour. He provides a commentary, like a reporter at a football match. Usually such commentary isn't really necessary. The TV audience are perfectly capable of seeing for themselves what is going on, but this is the way we do things. It is a completely normal part of human existence. After all, we are talking, rationalising animals, *zoon logikon*. In the case of Jayden, all kinds of interpretations are possible. Jayden might think that we are automatically reminded of such unlikely and incorrect scenarios as the one in which he is trying to ascertain why he has been feeling so weak lately. 'Oh no', his thoughts are communicating to us, 'I haven't been feeling ill at all. And I'm not a Zande either. That is not why I am inspecting this bird.' And then he continues out loud, 'I was wondering how it could have died.'

We do not only clarify and explain our behaviour in scenarios that may seem strange to others, are difficult to interpret or are easy to misunderstand. We do this all day long, in scenarios that are exceptional as well as in scenarios that are really quite ordinary. When a little boy leaves the living room, there's a good chance that he will tell you what he's going to do: 'I'm just off to the toilet.' And there's a good chance that you did the same when you were younger. We do not only comment on our own behaviour, but also on that of other people. And we do not always comment verbally, but also non-verbally. Sybille thinks she can see from my glance at the trees that I would rather go left, past the Birdman. We all know how you can show something like that by the way you glance at something. And we also all know how we can see that in the way other people glance at things. It is often such a subtle and beautiful process, the social interaction in which I use my body language to show Sybille what I want and she in turn shows me that she has understood: 'If you like, we could also go left to the turtle beach, past the Birdman.'

Of course sometimes things go wrong. I was in fact just looking at the way the sunlight fell on the leaves of the trees, and I had no ulterior motive. A glance like that is easily misconstrued, and this is why it is important that we also talk. Sybille tells me what she reads into my glance, and I give her my own interpretation. And we adjust our interpretations exceedingly smoothly, also in our body language, as in how you reacted when you arrived on Endoxa and took Esteban's hand to step onto the jetty. In the same way as we automatically adjust our behaviour to each other, we also adjust our words, our comments and our interpretation of the scenario in which we are engaged.[1] How this works and what is involved may be explained clearly with a story in which things go wrong.

This is a story that I once heard about a couple who found out about an implicit misunderstanding during the first week of their marriage. The story is set in the early 1960s. The wife had learnt from her mother that she should cook enough food, so that after the meal there was always something left over. For the wife, this was a matter of course, a sign of hospitality. There should always be more than enough for everybody. By contrast, the husband had learnt from his parents that you should always eat everything up. If you left some food in the dishes, the message was that the food had not been very tasty. After the wedding, their life with one another started: for the first time they lived together in their own home. And of course the wife cooked dinner, as that was the way things were in those days. She cooked a tasty meal and her husband ate everything. It was really good, he added, thinking his comment was understandable but also a little bit superfluous. After all, he had eaten everything, so his wife could clearly see how much he had liked the food. But what his wife saw was that she had not prepared enough. After all, everything had been eaten up. So the next day she prepared a little bit more. And again her husband ate everything. 'Wonderful', he said. 'Still not enough', she thought. This process was repeated during the next few days. And at the end of the week, the wife had cooked enough for a whole regiment. This time the husband was finally unable to eat everything. It just wasn't doable. It was really much too much. 'What's wrong here?' he wondered. He didn't understand it. And neither did she.

So what do you do then? Of course, the ancient Greeks already knew that this is the point that you start talking, rationalising your behaviour, interpreting the other person's behaviour and putting this behaviour within the context of the intended outcome. This is one of the obvious meanings of the verb 'humaning': clarifying what you are doing and asking the other person for such a clarification:

— Why have you been cooking more every day?

— I thought I hadn't cooked enough, that you were still hungry, because you had eaten it all. Why do you eat everything?

— I wanted to show you how much I liked your food.

Misunderstandings like these are quickly and easily resolved once you talk about them in the right manner. This is possible if we use our common sense, which is something that you will have to do a great deal on Endoxa. And as I explained in the previous chapter, you will notice that you start using your common sense as soon as you think about the entitlements and obligations that are implied by your expectations. These entitlements and obligations implicitly appeal to three different quality criteria for good humaning: understandability, efficiency and goodness. People want their behaviour to be understandable, efficient and good. I will discuss these quality criteria individually, but I will also show you how they are interrelated and interdependent. In addition I will show you how there are always different perspectives involved in the entitlements and obligations implied by our mutual expectations.

3. Understandable

People want their behaviour to be understandable, for themselves and for others. People want what they do to be part of a story that you can follow. This phrasing is telling: a story that you can follow. People live in scenarios, in situations in which something is going on, in which one thing logically follows from another. If people do something in a particular situation, if they act, they are responsible for a specific sequence of events. It is by their actions that the sequence is realised. And then they want you to understand them, to follow them, so that you understand why their actions were a logical reaction to what was happening. This understanding requires selective attention and a storyline. Not everything needs to be said; not everything can be said. You can leave out the details, but you do need the reference points that give the story its logical coherence.

If somebody enters a scenario halfway and has thus actually missed the beginning, people feel naturally obliged to offer some extra clarification. They explain why they are doing what they are doing, and why what they are doing is suitable behaviour at the time: 'I was wondering how it could have died.'

Have another good look at Jayden's explanation, and notice how naturally this explanation comes and what reference points he automatically focuses on. Jayden bends over to look at the dead bird that is lying at his feet, cut open. There is a reason why he does this; something is going on. Jayden knows what is going on, and he thinks that I do not. This is why Jayden thinks I need an explanation; otherwise I might start imagining things, making up a story about what is going on, and perhaps coming to the conclusion that Jayden is not quite sane. People do not usually look intently at the innards of dead birds. I know that, and so does Jayden. And Jayden also knows that I know, at least it is what he assumes and that is normal too. This is why he provides his explanation, 'I was wondering how it could have died.' And then he goes on. Notice how naturally he explores and discloses the path of his own behaviour. 'I found it just like that. An animal? Poison?' And then he offers an extra reason, apparently to explain why he has been looking at these innards for so long: 'I'm not sure yet what it could have been.'

These are the kind of stories that we tell each other about our behaviour. We do this against a backdrop of shared knowledge: an extensive, complex and detailed background. There is no need to pay attention to most of what you find in that background, it just needs to be in place so that we can follow the story. A background of shared knowledge excludes any alternative stories. Jayden is not a member of the Azande tribe, and of course he's not wondering whether he could hide a diamond inside that bird. The folk physics and folk psychology are present in the background, together with all kinds of more specific and local presuppositions about the things and people that play a role in this particular scenario.

This is how we continually estimate the understandability of what is happening to us. Understandability is then, despite all these possible details, a rather light and simple quality criterion. After all, so many different things can be understandable as long as you clarify them, as long as you use your selective attention to create a common thread in this scenario. This understandability is supported by two items: selective attention and a main storyline. Understandability is a narrative criterion, the quality criterion that revolves around the human ability to tell and follow stories.[2] It is a light and simple criterion because people are fantastic storytellers. Despite its implausibility, it is not at all difficult to understand the story in which Jayden had indeed found a diamond and was wondering whether he should hide it inside the dead bird.

This is why understandability is not the only quality criterion that we use when we rationalise our behaviour. It is, however, the most superficial criterion, with which we can easily outline the contours of a common world, a world in which we can relate to other people without having to enter into a lasting relationship with them. Understandability is an easy criterion for superficial contact. If my contact with other people lasts longer and becomes more intense, if I work together with them and develop a special relationship as a friend or colleague, understandability will no longer be enough. Then I will need more.

4. Efficient

In order to follow what somebody else is doing, it is enough for me to have a story that makes this behaviour understandable. But if I do not only want to be able to follow this person, but also want to work together with him, I need more than just understandability. I need to be able to count on him. I need to think that his actions are efficient. In this context, I use the word efficient as an equivalent of what philosophers tend to call 'instrumental rationality'. Basically, the idea is as follows: a person's behaviour is efficient if it serves as a proper means in a goal-driven plan. This efficiency is necessary for working together, because cooperation implies a common goal. This can be a minor goal, such as Esteban reaching for your hand and helping you onto the jetty. You will only manage to do so if you catch his hand and step out of the boat. It is as simple as that: a very elementary form of coordination in which you both have a goal, namely that you find yourself on the island of Endoxa. You may have all kinds of additional goals, goals in which grabbing each other's hands in the right manner is only a means of secondary importance. But to coordinate this one joint action, you need Esteban's efficiency (as he needs yours), for example to determine together how long you need to hold each other's hand for. Think about the jokey grandfather who shakes his grandson's hand, doesn't let go of it but in feigned shock asks the child to please let go of his hand. Always fun to watch.

Efficiency, as a sign of instrumental rationality, has been a typical feature of human existence since the days of Aristotle. That is: we are *zoon logikon*, sometimes translated (not completely correctly) as 'rational animals'. People don't just do things, but they explain why they are doing them. In other words, people have reasons for their actions. Ever since the ancient Greeks, these reasons have been associated with our intellect, with controlling our passions and emotions, with being cool and calculating, with conquering our impulses, and with choosing long-term goals over immediate and short-lived gratification.

In fact, the contrast between emotions and intellect is more or less secondary to my use of efficiency as one of the quality criteria for humaning. For me, efficiency only relates to the extent to which behaviour contributes to the execution of a plan. Sybille wants to take me to the turtle beach to eat something there, together with Anthony. We could go to the left, past the Birdman, or we could go straight ahead past the campfire. Straight ahead is shorter and there is more shade, but if we go left I will see more of the island, including the Birdman and the ruins of the Old Fort. There is something to be said for both options. Both routes are understandable. Both routes can also be efficient, but that depends on our plan. If we want to get to the turtle beach as quickly as possible, then going straight ahead is the only option that is efficient. However, if we want to see as much of Endoxa as possible, it is more efficient to go left past the Birdman, unless the time we gain when we go straight ahead contributes to our goal of seeing a great deal of the island. Of course this type of instrumental rationality appeals to our intellect. And it also requires control of our impulses. After all, to be efficient, you need to be able to think of a plan and execute that plan. You don't want to be distracted by what you happen to find in front of you. But being sensible, deliberate and thoughtful are not the most striking characteristics of efficiency.

The most interesting aspect of efficiency as a quality criterion for humaning is the role it plays in our interpretation of understandability. To make our own and each other's behaviour understandable, we implicitly make use of presuppositions and assumptions about the efficiency of our behaviour. And although understandability is a lighter criterion, it is supported by presuppositions of a heavier calibre, namely presuppositions about the suitability of certain behaviour in the light of an underlying, implicitly approved plan.[3] What I mean is the following: if Sybille asks Anthony whether he has forgotten what he wants to take from the cool box, the understandability of that question is supported by the efficiency of a plan that we as listener tacitly add to the background of the story. After all, you look inside a cool box so that you can take something out. That is not the way you look inside a dead bird, and this is why Jayden felt compelled to explain why he was doing this. Sybille puts her arm around Dituju and leads her along to the turtle beach. This is understandable because we can see the efficiency of Sybille's actions: Dituju needs comforting and you can offer someone comfort by keeping them company and by giving them something to eat. And the role of Sybille's own ideas about the route to the turtle beach becomes clear when she asks me if I'd rather walk past the Birdman. It is as if she thinks I cannot simply be looking at the way the light falls on the leaves, but that I must have an ulterior motive.

People are rather lazy when making explicit all these presuppositions about each other's efficiency. Someone does something and we automatically attribute an intention to that person, so that the behaviour of this person can be part of a world that is understandable to us. People carry out their plans in such a world, although we are often not at all sure whether those people really have these intentions and whether they are really carrying out the plans we think they have. We are not really interested. We have a blind faith in our assumptions, and to be fair, usually this doesn't cause any problems, usually our assumptions simply keep the world neatly arranged so that we can carry out our own plans and live our own lives. Perhaps the following situation is familiar to you. During the interval at the theatre you simply follow everybody else, without giving it any thought, convinced that just like you these people will be on their way to the foyer. Then you suddenly realise that you have followed them to the men's room.

Interesting. What has gone wrong here? Is these people's behaviour impossible to understand? No, of course not, that would only be the case if they had actually wanted to go to the foyer but had not paid any attention to their route. Like you hadn't. It is your behaviour that cannot be understood, not theirs. It would only be impossible to understand their behaviour if your specific idea about what people should do during the interval at the theatre (walking to the foyer for a glass of wine) were the standard. This brings me to an important point. People try to show understandable behaviour, and this is why they often explain what they are doing, verbally and also non-verbally. This explanation clarifies their behaviour. It places their behaviour in a wider context of efficiency, in a plan that they carry out. Much of that plan remains presupposed and in the background. You might perhaps even say that plans tend to remain hidden in the background, made invisible by what actually is said. After all, the background is the background, the thing that people do not pay attention to, the thing that remains implicit to all people concerned. But that is exactly why it is not explicitly clear to what extent this background is one and the same background for everybody concerned.

This adds interesting dynamics to our interpretation of the extent to which our behaviour meets the quality criteria of humaning.

5. Good

One essential aspect of the background often remains hidden and unnamed, just when you would like to explain the efficiency of your behaviour by putting it in the context of a plan. This aspect relates to the following question: is the plan good? Is it correct to have this plan: is it suitable, fitting, reasonable? Jayden asks whether we have come to visit the ruins. Sybille answers him briefly, intelligibly and efficiently: 'Well, actually we are on our way to the turtle beach. Anthony is there and has something to eat for us.' This is just a passing remark. Still, Sybille makes a great deal clear. She makes our behaviour understandable. We are carrying out a plan. We are hungry. We want to eat something. Anthony has made something to eat. He is on the turtle beach. That is why we are walking there.

This is a clear story for the listener, but only because of the pre-supposition that at this moment, in this situation, this is a good plan to carry out. Within the context of the plan it is understandable that we do not accept Jayden's invitation. If a listener thinks this is a clear and understandable story, it is because he, the listener, implicitly accepts that it is a good plan for me and Sybille, at this moment, in this situation. This listener would find it highly unlikely that the following is actually the case:

The Apate

Jayden hasn't eaten for 2 weeks and has been looking for something edible for days. He has often found some dead animals, but they have all been poisoned. Jayden is being blackmailed by a secret society called The Apate, and this society wields the power on Endoxa. Jayden will only be allowed to come and eat on the turtle beach after he has shown fifty visitors the ruins of the Old Fort. Sybille is a member of the Apate and she knows that Jayden has already given forty-nine people a tour of the fort.

In this case, we are the listeners, you and I: me because I am Sybille's companion and you because you are the reader of this book. But how do you know that there isn't a secret society called The Apate on the island? Personally, I had no idea.

It says a great deal about the experimental, exploring, investigative and also risky character of humaning that I can end up talking about the Apate when I thought I would simply be enjoying a visit to Endoxa, without a worry in the world. Please let me explain this notion step by step.

I am walking with Sybille across Endoxa. Sybille understands my behaviour and attributes a certain efficiency to it. The reverse is also true: I understand her behaviour and also attribute a certain efficiency to it. We behave in different ways, but these ways correspond to one another because we both presuppose that our behaviour derives from one and the same plan. However, the scenario *The Apate* shows us that this is not necessarily true. There may well be plans that are different but still seem to correspond: in other words, different plans that under certain circumstances may result in identical effects. There are a variety of ways in which plans that are different can be made to correspond. In the scenario *The Apate* it is Sybille who makes this happen. She has an overarching plan that involves my plan, in a way that she controls.

Such a one-sided construction is necessary in the story of *The Apate* because Sybille certainly knows that I would never cooperate if I knew what her plan was. Sybille's plan is in fact an evil plan, and it becomes clear from this plan that her behaviour does not meet the quality criterion 'goodness'. It is not good to blackmail Jayden and to starve him. If that is what Sybille is doing, she is bad at humaning. Sybille needs to prevent me from drawing that particular conclusion because then I would no longer be capable of working together with her. After all, I would no longer be able to see how our behaviour could be efficient, nor how it could be understandable. The behaviour that Sybille and I display in the story of *The Apate* can only correspond if she deceives me, if she manages to make me believe that her behaviour is good.

It is not necessary at all for Sybille to do her best to convince me that her behaviour is morally good, either explicitly or in a narrow sense. Of course not, that would only look suspicious. But if she wants me to work together with her (and this cooperation can be as superficial as walking somewhere together), for me her behaviour will have to be understandable and efficient, and certainly not dubious. It should not be possible for me to see any harm in her behaviour. I would simply take for granted that the goodness of her behaviour is present in the background, as would be expected. This goodness is necessary for me to be able to follow the efficiency and understandability of her behaviour in a manner that is non-problematic and self-evident. This understandability is based on a presupposed efficiency, just like the efficiency is based on a presupposed goodness.

All that Sybille needs in our common background is the – to me self-evident – goodness of our behaviour. In this there is a key role for my perspective on our plan (and thus the understandability and efficiency of our behaviour), because it is in terms of my perspective that our plan is in fact my plan, a plan that Sybille can use in her more elaborate plan. In my plan the goodness of our behaviour is not an issue at all, not a point of attention. That is the way Sybille likes it and needs it to be in order to manipulate me into cooperating with her. For me the goodness of our behaviour goes without saying, but not for Sybille. And this is related to the distinction between the three different quality criteria for humaning: understandability, efficiency and goodness. Sybille is aware of this distinction and it is important for her that she distinguishes these three criteria. Sybille has to persuade me to come with her, and this is why she has to obscure the immorality of our behaviour and to pretend it is in fact efficient.

In short, this presupposed goodness is not only expected in the background but also necessary to make working together possible. The role of this goodness can be explained further by what happens during a football match. When the football players enter the field, they obviously have different plans. First of all, one team will do their utmost to make the other team lose, and the reverse is clearly true as well. So there are already two contrary plans visible on the football pitch. In a harmonious manner, these two plans are part of an overarching plan, a joint plan of all the twenty-two players to play a football match. This is where a football match clearly differs from the way in which Sybille is trying to manipulate me on Endoxa, and in which she includes my plan in her overarching plan that she has to keep secret from me. Of course, both teams also try hard to keep their strategic plans hidden from the others, but that happens within the limits of the overarching plan that they have both subscribed to.

This plan makes it possible to qualify all kinds of behaviour as understandable, efficient and good. Understandable: football players are not allowed to pick up the ball and hide it under their shirts. Nor are they allowed to carry guns to intimidate their opponents. They are not allowed to have two goalkeepers in the field at the same time and so on. Efficient: football players respect the decisions the referee makes, they do not make unnecessary fouls, they divide their energy in such a way that they can still run and shoot in the final minutes of the match, they do not take off their shirts when they have scored a goal and so on. Good: football players know that they need to be good sportsmen and play fair, that every team needs to be a team, that the opponents are only opponents in a sports sense for the duration of this match and so on.

A specific goodness is also presupposed in the personal plan with which every football player enters the pitch. As a player you are committed to your team. It is clear that you will do your utmost, that you will be a good team player, and so on. But it is also presupposed that as a football player you are committed to the game of football as a sports battle. Usually all this remains implicit, but suppose that your immediate opponent went to school with your wife and bullied her for years . . . Then you would have to wonder whether under these circumstances you can remain loyal to the goodness that is presupposed in the game of football: that you are a good sport and will behave in a sportsmanlike manner. Countless moral deliberations then suddenly emerge, but such things of course only happen if there is a concrete reason. Would it be conceivable for a Palestinian to play in a football match against Israel? Is it possible to play football if you have slept badly for weeks? If you are gay, and you're afraid that your homophobic team mates will find out? And so on.

The important point of this elaboration on humaning on Endoxa is that for all our behaviour, whether it is playing football, singing, walking or humaning, there are three quality criteria that tacitly function in the background. I have discussed three aspects that can make the specific content of these criteria ambivalent and ambiguous. First, the criteria are – by the expectations of the people involved – related to all kinds of different plans, perspectives and orientations. Second, these plans, perspectives and orientations can be interwoven in a manner that may or may not be open, and that may or may not be contradictory. And third, these criteria are related in an asymmetrical manner: goodness limits efficiency, and efficiency limits understandability. Two special relationships are involved that are crucial for us humans and our humaning: cooperation and identification.

6. Humaning

Suppose you see someone behaving in a way that is completely incomprehensible. It is actually rather weird to see someone inspecting the insides of a bird, like Jayden did, but think of something that is even weirder. Like a Monty Python sketch. Or imagine seeing Sumalee on the turtle beach, showing a piece of pumice how it can jump over a wire that she has fastened one foot high. You might think, oh well, why not? What is it to do with me? But indifferent or not, your imagination will have been activated because something that you see and don't understand automatically creates cognitive dissonance. We human beings like our world to be well-organised, understandable and compatible with our folk psychology and folk physics. These are necessary conditions for humaning.

The most intriguing aspect about understandability as a quality criterion is that it is interpersonal. It is a characteristic that we can ascribe to behaviour because of the cognitive efforts made by our imagination. Behaviour is understandable if we experience it as understandable; what we do not find understandable we can make understandable, so that eventually it is understandable. In this way, the power of our imagination is also its limitation. Partly, understandability is in the eye of the beholder: it is a function of our perspective, of our view of the world, and our real challenge is to get other people to adopt our perspective, and for us to take up their perspective.

Behaviour that we understand makes us calm, gives us cognitive peace of mind. If other people behave in an understandable manner, we find ourselves in a world that is kind to us, and does not exhaust our imagination, so that we can use our cognitive resources for other things. If we do not need these other people, if we merely observe them and do not have to work together, it is enough for us that their behaviour is understandable. If that is the case, these people do not need any further attention. They are then the background to our world, together with all the other things that do not surprise us. They derive their understandability from our background, our expectations, our folk psychology, our folk physics and our plans.

But if you want to work together, understandability is not enough. If you take Esteban's hand to step onto the jetty on Endoxa, you must think his behaviour is efficient, and he must think the same of your behaviour. You both need to subscribe to the same plan, which may be as small as stepping onto the jetty in the right manner or walking together to the turtle beach, but it may also be as complex and fiendishly ambiguous as Jayden's attempts to inform me about the dirty tricks the Apate play, so that I could become his ally in trying to end the mafia practices of Sybille and her evil secret society.

In an interesting way, cooperation is linked to an even more intimate relationship, namely identification. People sometimes identify with each other or, more specifically, identify with a certain type of person and assume that significant other people identify with that same type of person.[4] This can make a greater or lesser demand on the people concerned, but it certainly requires more than just efficiency. For identification you need, at least within the limits of the scenario in which identification plays a role, a shared value orientation, a shared idea of what makes behaviour good. This is a well-known phenomenon in the context of professional football and its permanently on-going transfer circus. A football player who has been sold to a different team and then has to play against his old team finds himself in a situation in which he has to emphasise his new identification without denying his old identification too clearly.

In *The Apate*, both Sybille and Jayden might try and persuade me to identify with them in their plan. That would require me to think that their plan is a good plan. If Sybille has no reason to suspect that I will dissociate myself from her secret society, she will not try to manipulate me but could inform me light-heartedly about the mutual relationship between her and Jayden. There must have been similar cases on the beaches of Ghana in the seventeenth century, when Portuguese tradesmen met Dutchmen and Britons, and everybody knew full well what they were all there for: shipping strong slaves to the South American plantations.

In the scenario of *The Apate*, Jayden might deduce from Sybille's behaviour that I would sooner be inclined to support him than her. After all, the manner in which she answers Jayden betrays that she hasn't told me anything about her evil schemes. Apparently she wants to keep up the delusion that we are three people of good will here, each with enough autonomy to carry out their own plans. As far as I know, I am just on Endoxa for today to see how things go here. This will give Jayden enough information to realise that there is no explicit identification between me and the evil Apate society. Apparently I have no idea what is going on and I have only implicitly identified with the general type of 'people of good will'. Jayden may suppose that this type of person has more affinity with him, with his hunger and his unequal struggle, than with the Apate extortionists. There are significant interests involved here and the struggle is extremely unequal, which makes it very difficult for Jayden to operate in *The Apate*.

I am glad I am not in his shoes. I am glad that I have simply made up the Apate (and Endoxa). I have done so to make something clear, namely that there is always a complex, structured background to humaning. In this background, three quality criteria are tacitly presupposed, which give people all sorts of ways of leaving each other alone, give them all sorts of plans that make cooperation possible, and give them all sorts of expectations about how normal 'people of good will' behave. Within this shared background, the expectations of each other's good will are the most deeply and most implicitly presupposed. As I have shown in this chapter, they modulate what appears before us as obviously efficient and, as a result, obviously understandable. The expectations regarding other people's good will are the last to become apparent because they are only articulated explicitly if there is tension because the processes of identification do not match. But that is not something that occurs easily. Usually, identification is a background phenomenon; it only comes to the fore if the people cooperating have plans that do not correspond.

However, cooperation can usually be realised, especially against the background of a general idea of 'people of good will'. Earlier we saw that we can produce understandability if we do not find it; similarly we can produce cooperation, even when it is unclear whether our plans correspond. You only have to think about what happens when you enter a baker's shop and you need to wait your turn. Together with the other waiting customers you spontaneously form a queue, as this produces the greatest cognitive peace of mind. Note that you only do this, and can only manage to do this together, if you tacitly presuppose in the background that it is good if you do not jump the queue, do not push, and respect the idea of 'first come, first served'. Interestingly, this moral background is so obvious that you hardly notice it and barely realise that it is the moral background that makes you interpret each other's behaviour as efficient and therefore understandable. This is what we all do at the baker's. We scan to see who was already inside before us and we try to put these people in a queue. And then we pay attention to whoever comes in after us and we put these people in a queue, too. Exactly between these two points, we all discover our own position, individually and jointly.

This moral background – the general idea about how people of good will do their humaning – is a self-evident fact, and it is also present on Endoxa. We need this background, always and everywhere, if we rationalise one another's behaviour, if we try to understand each other and if we try to cooperate with each other. I apologise that I needed such an evil example as *The Apate* to make this clear. Scenarios such as *The Apate*, however, do not cause the greatest problems for people who want to be humaning. Of course, there may well be indisputably evil secret societies that threaten the act of humaning. I do not want to sound naïve about that. But they can't just quickly be put together, and the tendency for organised malice is not very common. Problems involving everyday rationalising of our behaviour may sooner be expected from misunderstandings and incomprehension. If these become too large, you will need more guarantees than an obvious but implicit moral background. You will need something stronger. You will need an extensive version of common sense, more than just the expectations of folk psychology and folk physics, and more than just the skills to interpret our behaviour as understandable, efficient and good. In the next chapter I will explore two capacities that can reinforce our common sense in this particular field: trust and accommodation.

Notes

1 The fine details of such interactions have extensively been studied by ethnomethodologists such as Garfinkel (1967). See also Schutz (1967).

2 Narrative explanation is a major theme in David Velleman's work over the years. See e.g. Velleman (2009). For a detailed analysis of both the strengths and the weaknesses of understandability as a criterion of what I call good humaning, see his Velleman (2003).

3 The analysis of planning agency is the major theme of Michael Bratman's work. See Bratman (2014).

4 Much of what I have been arguing for in this chapter is reminiscent of general themes in the work of ethnomethodologists, such as Garfinkel (1967) and Goffman (1956), symbolic interactionists, such as Mead (1934) and Blumer (1969), and their pragmatist predecessor Dewey (1922).

4 Trust and accommodation

1. Before you go to sleep

You don't fall asleep immediately, which you rather like. So much has happened today, too much to just let go. There is so much to think about and figure out. It will be some time before it is all in balance again.

You ponder over the day's events. Esteban who carefully sought out your hand, squeezed it lightly and then held your hand for quite a while, caressing it softly. It was comforting, and also exciting. Of course that was the highlight of the day for you, but what happened before that . . .

Thebe's tears made quite an impression. She was so sad. She felt so embarrassed, betrayed, vulnerable; she was dreadfully upset. How could anybody do something like that? Katya had a nerve . . . And Thebe was also really disappointed in Ningak. Actually, it was all a bit like a TV drama. But why on earth hadn't she just put her name to the poem? She wasn't an adolescent now, was she? Anyway, it was all out in the open now.

Funny woman, Thebe. Keeping so very quiet that she was in love; such a romantic, yet timid person. She hadn't mentioned it to anyone; she must have been mulling it over all the time. You can just imagine that she must have been fantasising about Ningak, and that her fantasies had spun out of control.

Then there was the poem. Sweet is the first word that now pops up in your mind. Or is that too condescending? As if you're talking about a child. Well, actually it all seems rather like puppy love. An anonymous poem. Why would you do that? And what would the poem have been like?

Sweet, of course, but also a bit silly. Immature.

And then there is Katya. You hadn't imagined that she was so sly. But then again, it may all have been a coincidence. It is not impossible. An accident of convenience.

Perhaps this is what happened: Ningak got the poem and was touched, moved. Sounds plausible, and to be fair, it must feel good to receive a poem from an anonymous admirer. And perhaps that was just when Katya entered, seeing Ningak holding that paper. She likes him too, you know. It's impossible not to notice. So who knows, she may have started blushing, making Ningak feel awkward with the poem in his hand. He also went red and to save the situation he suddenly blurts out, Katya, is this yours? Did you send me this poem? So it may just have been a series of events that caused all these emotions. Simply bad luck.

You are uncertain what to believe. Ningak may of course have hoped that the poem was Katya's. Perhaps he has feelings for her. Or there may have been something in the poem that reminded him of her. If that is the case it is easy to understand Thebe's sorrow. Unrequited love is always such a sad thing. Terrible.

You lie still for a bit; your thoughts seem to be settling.

Then the memory of Esteban's hand comes back. Here you are in the dark, smiling, glowing a little. How did it happen, this thing with his hand? What made him do it? How did he know that it was alright, that it was exactly the right thing to do at that moment: his hand holding yours and stroking it softly.

Clearly the emotions shown by Thebe, Ningak, Katya, and Dick – Dick's anger – had opened up your feelings as well. You wanted to be touched, and Esteban sensed this.

You were amazed by Dick's anger. Actually, he had had nothing to do with the whole situation. What could have triggered him? Did he feel it was directed at him? But how? He was absolutely furious with Katya. He felt that it was just not done. In hindsight Dick had been rather ridiculous; the reproaches he made were rather pathetic. He was talking about the soul of the poet, and how Katya had crushed it, trampled Thebe's soul. He had even talked about the soul of Endoxa, and how it wouldn't surprise him if something nasty like this would have detrimental effects in the long run. That the people on the island would all be more on their guard, that you would no longer dare to show your vulnerable side to the other visitors on the island. That this was a serious blow to Endoxa as an open society.

Perhaps it was because of what he had said. Perhaps you and Esteban wanted to prove him wrong. Esteban was standing right next to you, almost leaning against you. You were both watching Dick speaking, Thebe crying, and Ningak feeling uncomfortable. Katya was trying not to show her emotions, and had a distant expression on her face. And then you felt how close Esteban was, and apparently so did he. He moved his hand, took yours, squeezed it lightly and then held it for quite a while. He softly stroked your hand, comforting. And it was really exciting. So intimate. Lovely.

2. Emotions

Human beings are sensitive creatures. When we start our lives, we are even hypersensitive. Completely off balance due to the intrusive force of birth, we start crying. And for a long time after that we are dominated by our primary emotions: sadness, fear, anger, curiosity, sympathy and joy.

In fact we are beautiful, complex, delicate beings who are sensitive to an almost inexhaustible range of stimuli. If we compare the sensitivity of us humans to our sophisticated computers, it is easy to see how poorly their sensitivity is developed: all they feel is the touch of the keyboard and the click of the mouse, that's all. And what about plants? Sunlight, water and some chemical interactions with their environment, that's about as far as their sensitivity goes. Animals offer a much wider variety: they are also sensitive to movement, scent, sound and vibrations. But what is also striking is that animals all have their own little niche, their relatively limited habitat in which they are at home, in which their reactions are only geared towards the stimuli that matter. Your hand moves slowly towards a mosquito and then you strike. In the bloodstain on the wall there is nothing that shows what actually happened.

Of course, in a way this is also true for human beings. We also have our habitats; after all, we are animals, and just like many animals we can hear the thunder rumbling in the distance, can distinguish subtle differences in colour, can notice movement in the background, and can sense it if there is something wrong. But we are also tremendously sensitive to a wide range of social and relational stimuli. Not only can we register the smallest nuances in a voice, detect changes in a person's attitude, and understand facial expressions: that is something that many domesticated animals can do, too. But we are also sensitive to what other people say, to their choice of words, and to the content of their message.

There has been a long tradition in Western philosophy in which the difference between man and beast is explained in terms of our capacity to think in abstract, logical and rational terms. Animals have to make do with body language, as this is the only language that they have access to. But according to this tradition, human beings actually speak, using words with content, with an abstract meaning. Human beings are linguistic creatures, who can give and accept emotionally detached, logically compelling reasons for their behaviour. Only human beings have access to a rational and intellectual world of meanings, symbols, ideas and reasons. It is characteristic of this tradition that it discerns a shrill contrast between the intellectual and the physical, as if there is no link between intellectual sensitivity and our emotions and our body, as if we are made up of two different parts: intellect and body, two parts that react to two different types of stimuli.

By now, this tradition seems to have run its course. Increasingly, emotions are being considered valuable. They appear to have their own rationality and are vitally important for our moral judgement (Cf. Damasio, 1994). They form the basis of our ability for practical reasoning and for the understanding that certain facts can serve as relevant motivation for our actions. They appear also to play a major role in our ability to deal with our environment. These days, emotions are hot.

Indeed, in this chapter I will join this topical, sentimentalist surge in favour of emotions. I will argue that emotions play a crucial role in our common sense, in our ability to do our humaning, and in our ability to adequately adapt to new situations. Common sense is based on emotions, on elementary, primitive emotions, and especially on their developed, refined transformations: emotions that contribute to behaviour that is understandable, efficient and good.[1] The fact that I am shocked when I unexpectedly stumble on a boa constrictor that I find hissing in my shed is good and sensible: this is a suitable emotional reaction to a dangerous situation. This shock is like a reflex, similar to the way we automatically pull back our hand from a heater that is much too hot. The reflex aspect of such an elementary fear does not detract from its suitability. It is a good thing that these kinds of primitive emotions are well engrained in our bodies. From an evolutionary perspective we probably owe them our existence.

But emotions are not only primitive and reflex-based. Many of our emotions have developed further, are refined, have been transformed from primitive reactions to responses enriched by knowledge and experience. Of course, they still maintain the characteristics of emotions. In the words of Jesse Prinz, they are 'embodied appraisals': evaluations of a scenario that express themselves physically (Prinz, 2004). Emotions shape the way in which we as living creatures strive to ensure our own survival. They put us in a state of readiness. They prepare us physically for an adequate reaction to a scenario. Refined emotions, emotions enriched by knowledge and experience, do this in the same way as primitive emotions, although obviously this feels different. If I see a drunk walking towards me with a broken wine bottle, I will experience a completely different reaction to that of an experienced police officer for whom this is all in a day's work.

The development of refined emotions is supported by two complementary abilities of human kindness: trust and accommodation. At least, that is what I will argue in the remainder of this chapter. Trust and accommodation are important components of our common sense; they are important social skills that are related not to a particular culture, but to human beings living together in general. This is why trust and accommodation can flourish on the island of Endoxa, where you need to do your humaning without access to expert advice.

3. Entrusting

So what was going on? Why was Dick so angry with Katya, who pretended to Ningak that she had written Thebe's poem? Why did he rant on about the crushed soul of Endoxa and the effects of this disgraceful deceit? Of course, these are rather specific questions but they represent the kind of questions that we can now deal with, building on from my argumentation in the previous two chapters. Apparently Dick had certain expectations about how Katya, Ningak, Thebe and others should behave on the island of Endoxa, and he felt that these expectations had been thwarted, especially by Katya. This is why he spoke in such general terms: he was trying to make clear that Katya's behaviour would have a detrimental effect on the understandability, efficiency and goodness of humaning on Endoxa. It seems like a standard example of what we do when we rationalise behaviour. But there are some follow-up questions that focus on the emotional side of our behaviour, the side that implicitly played a role in the previous two chapters but which has not yet become a theme. Why did Dick become angry? Why Dick? And why did Esteban react to this by taking your hand, at least if that is the correct interpretation?

If we don't pay attention to the emotional side of our behaviour, rationalising will not lead to anything. It is as if you are leaving something out that is absolutely important. Or worse, as if you are ignoring something crucial. Still, this works both ways: does Dick's speech fail to pay attention to his own anger? Or is he ignoring the emotions of the others, such as Ningak and Thebe? Should Dick have been more understanding of Katya's envy, Ningak's timidity and Thebe's sadness, or should he simply have realised that he was angry himself? These are all important, of course. All of these emotions merit our attention, both the emotions that provoke rationalisation (in this case Dick's anger) and the emotions that are marginalised in this rationalisation (in this case Katya's and Ningak's emotions) or have become enlarged (Thebe's sadness). It is to validate these emotions that I would like to focus on the role of trust, both in our own behaviour and in the rationalisation of that behaviour.[2]

Trust starts with entrusting, leaving a task to somebody else. Every football team needs a goalkeeper; this role has to be entrusted to a team member. In a similar way you entrust your baker with baking your bread, the bank with taking care of your money, and the clothes shop with providing you with jeans that are comfortable, not too tight and not too baggy. From beginning to end, human coexistence revolves around entrusting tasks to other people, and you expect or hope that these people will carry out their tasks adequately. I am using the rather formal word 'task' here to describe behaviour that a person has the informal obligation to exhibit, while other people have the informal entitlement to expect that behaviour. In that sense, Dick has apparently entrusted Katya and Ningak with a task, namely the task of interacting with other people in a respectful manner; in his own words: the task not to crush the soul of Endoxa.

Sometimes entrusting takes place in an explicit manner. At a meeting someone is appointed the task of minutes taker, at a party one person will be in charge of the snacks and someone else of the music, and in a football team everybody has a clear task, too. But even in a football team, which is so explicitly organised, there are all kinds of tasks that need to be implicitly and extemporaneously entrusted to different team members: during the game, in the dressing room, at the press conference, in the players' bus and so on. This happens spontaneously and you can see this in the behaviour of the people concerned, in the way in which people feel entitled to expect certain behaviour from other people. For example, Dick has never explicitly entrusted Katya and Ningak with the task of not crushing Endoxa's soul. But he has developed a certain expectation about the behaviour of these two people and the rest of you on the island. That is a matter of folk physics and folk psychology, which we discussed in Chapter 2. That is also where I argued that expectations based on psychology bring in their wake entitlements and obligations. People may be aware of those entitlements and obligations; they may explicitly think they have these entitlements and obligations, but usually this happens implicitly, simply by having expectations. And that is especially true if the task that is entrusted to others is simply what I call – together with Kuitert and Aristotle – humaning. We simply expect people to do their humaning, always and everywhere, wherever they are and whatever else they are doing.

At this point a whole series of instructive emotions present themselves, emotions related to the interesting fact that if we want to start humaning, we need humaning fellow humans. When we are humaning, we always do so with the expectation that our fellow human beings will also be humaning. In everything we do, we trust that the people that we contact will be humaning, too. And we don't mean that they are only willing to do some offhand humaning, that their humaning would not exceed the C- level; we expect them to do their best, to go for it, to do their humaning with enthusiasm, determination, resolution and to the best of their ability. We entrust them with that task. We have no choice, because we will not be able to do any humaning at all without their commitment. And this is where these emotions come from.

Take a look at Dick. He is angry, conspicuously and openly angry. It is clear that he thinks he has a right to be angry. He does not doubt that his expectations are correct. He thinks that it is completely clear to Katya and Ningak that they should be humaning. For him that means that they do not betray Thebe, as that would be outrageous. Completely below par. They have clearly abused Dick's trust.

Now let's have a look at Thebe. Clearly she has been affected emotionally. She is sad, heartbroken. She had also assumed that Ningak and Katya would be humaning. And for her this means that her love poem would be in good hands with Ningak, that he would deal with it in a respectful and discreet manner. And Katya would have kept out of it, as this is what people do if they are not explicitly invited to interfere in an intimate relationship.

Esteban's reaction – holding your hand – was also mostly emotional, and so was yours by the way. In bed at night you adopt a more contemplative attitude. Who knows, Dick, Katya, Thebe and Ningak may also be lying awake. What would their thoughts be about what had happened? Will they realise that all things revolved around entrusting, around the realisation that we humans are at the mercy of other people's ability to do their humaning properly? Because that is what you realise.

This is what you are thinking. When you come into the world as a baby, a small bundle of a human being, you are completely at the mercy of the people in your environment. You can do hardly anything, only the most basic things: breathing, living, humaning. But the way in which you do your humaning as a baby is characterised by moving, touching and almost total surrender. To do your humaning, you are utterly dependent on other people's actions and good care. Your life is entirely in the hands of other people. Your complete existence is a matter of trust. You are nothing more than trust. You *are living* trust. In other words, your activity is no more than simply being passive, being receptive, being emotionally touched. Your entire existence is a matter of being passionate.

These two words belong together: passive and passionate. The first word emphasises the passive side of your existence, the fact that you are an object to whom things happen. The second word, however, emphasises the active side of your existence, the fact that you try to reach your aims with passion. A baby is passionate in his humaning: with total surrender he trusts the humaning of his fellow humans. Both words, passive and passionate are related to the word passion, a concept that is closely related to words like sentiment and emotion. These two words are used well in Jesse Prinz's characterisation of what he calls emotion: 'embodied appraisal'; we saw this above. As a baby you cry. All your activity is screaming: you passionately make clear that you are there, that you matter, that you are of immeasurable value, and this value presents itself as total surrender, complete trust.

You can say of yourself as a baby that you were completely entrusted to your parents. Entrusting is also a verb, but it is more a state than a process, a fact rather than an activity. It seems to be something that you *are* rather than something that you *do*. Still, this is what the verb humaning was for you as a baby: it's what you did by breathing, by living. The whole of you was entrusting. It was your situation, your whole existence. It made you extremely vulnerable, and your life extremely emotional.

This is what you concluded as you were still lying awake, mulling over the things that had happened that day, moved by Esteban's touch. The verb humaning is the same as entrusting, simply being like a baby: vulnerable, emotional, at the mercy of others. But by and by, this entrusting changes its image, because you develop expectations, because you become more at home in folk physics and folk psychology, because you learn to speak and to rationalise, because you can give your own contribution to the way in which we live together, and you start to value the understandability, efficiency and goodness in that. Entrusting, you think, becomes trusting: an adult way of dealing with each other. Two complementary aspects meet in this: reliability and vulnerability. And emotions, you think, pop up if the balance between reliability and vulnerability is disturbed. If your vulnerability unexpectedly becomes dominant, as happened to Dick and Thebe, you may become angry, sad, or afraid. If the reliability of another person unexpectedly becomes dominant, for example Esteban's reliability, you become happy, touched. Sounds rather good, you think, and contemplating Esteban's touch you fall asleep with a big smile on your face.

4. Reliability

That is how we all start our lives, completely dependent on other people caring for us. Evidently, that has gone well, or you wouldn't be here to read this.

But still . . . It is necessary to add some nuance. I know that children are exceedingly flexible, shockingly naïve in their trust, and terribly skilled at adapting. Who knows, you may have indeed made it, survived, but you may have emerged from your younger years with obvious scars. Your trust may have been abused. It may have become abundantly clear that you couldn't entrust your parents with your life. They may have turned out to be completely unreliable. You may even think that they should never have had children. That is possible, even though it is an exceptionally harsh verdict, one that we usually only give about other people's parents. It is an extreme variation of an idea that we can all agree with, to a greater or lesser degree: you cannot trust everybody, the reliability of our fellow human beings is sometimes seriously lacking. This leaves traces in each one of us, in the emotions that show that we are losing our emotional balance. It would make an enormous difference if we could influence other people's reliability.

I am not sure how many people think this, and I am not sure how many people have a basic attitude characterised by insecurity, or, in other words, by the idea that you first need to have a good impression of someone's reliability before you want to or dare to trust this person. But once you have grown enough to actively and explicitly entrust other people with something, there is still the associated realisation that doing this is related to the other person's reliability. At first sight this presents us with two opportunities: either you begin by entrusting somebody with something because you yourself dare to carry the responsibility for that, or you begin by determining whether someone is reliable because you want to be certain that he or she can carry the responsibility involved. And if this is (or seems to be) the choice you need to make, then there is something to be said for the second, safe option: first try to find out how reliable the other person is.

However, this is a dead-end street, a hopelessly incoherent undertaking, and below I will show you why. It is behaviour that has been growing in popularity over the past few decades. At the beginning of the twenty-first century, we have developed a true passion for determining other people's reliability. It is no longer trust that we are dealing with, but security, and we have a modern strategy for this: we have *experts*, people with knowledge and qualifications, people to whom you can entrust all your affairs. Of course you can try to do everything yourself: bake your own bread, fix your own shoes, unblock your own sink, grow your own vegetables, brew your own beer, blow your own glass, upgrade your own mobile phone and install the latest version of Firefox all by yourself. But everybody knows this won't work. You have to leave something, quite a lot of things actually, to other people; you need to entrust these things to others. And that is why we have experts, specialists who have mastered a certain profession, and this is why they are reliable in a specific field. They know what they are doing, and as such are completely different from the amateurs and cowboys who try to impress you but who are not reliable because they don't have the expertise.

And the experts hold the key to trust – if you are naïve. Experts are by definition reliable, and you can confidently leave your affairs to reliable people, to experts. There's only one problem: how can you know who is an expert in a certain field if you are no expert in that field yourself? And, of course all this is *only* about the fields in which you are not an expert yourself, since it is only in these fields that you need a reliable other person. Still, the fact that these are exactly the fields in which you are not an expert yourself makes you incapable of judging whether another person is. Anybody can *say* they are an expert.

And thus we stumble across a modern-day version of Juvenal's classic problem: who will guard the guardians? Sadly enough we seem to forget nowadays that Juvenal was a cynical satirist who did not want to discuss a serious problem, but who was criticising decadent excess. Even though we take the problem seriously, we do not really understand it. We look in the wrong places and thus create a new excess, not a decadent one, but in the end a much more wretched one: excess of mistrust. Because we have a destructive answer to the question who guards the reliability of experts: *reliability experts*. En masse these people have entered education and academia: accountants, supervisory bodies and external review panels that carefully determine that a qualification correctly represents that which it is meant to represent. Our experts are reliable. That is the truth: after all, they have a certificate, a guaranteed certificate with the stamp of approval from the training centre. And this training centre itself also has a stamp of approval: it has been accredited; it has received a certificate as well. And yes, the expert members of the review panel have also obtained their qualifications. It seems to go on indefinitely: every expert has a certificate, every certificate has a stamp of approval, every stamp of approval has been accredited, every accreditation needs an expert, with a certificate, with a stamp of approval, and so on.

It may not be completely fair to lash out like this against this situation. After all there is a serious relationship between reliability and expertise. For example, let's consider a mechanic. When I hear a funny grinding noise somewhere around my car's right rear wheel, my first reaction is to turn up the volume on the car radio. Granted, not the cleverest response. But then again, this is not my field of expertise. When I take the car to the garage, I expect the mechanic to recognise the grinding noise, to know what to do, and to actually do this. I expect him to have learnt this, to have gained objective knowledge about the way a car works and as a result he is programmed as it were to discover the cause of the funny noise. I expect him to have obtained knowledge and experience, which enable him to react to what he observes in a systematic, accurate and adequate way. That makes him reliable. Like a thermostat, if you like. We say that a thermostat is reliable if it registers the room temperature and then does what it is supposed to do: switch the heating on or off.

Like an investment advisor is reliable: he knows what the figures mean, he knows what he should do in response, and he actually does it. And like a plumber is reliable: he knows what causes that awful stench, he knows what he should do in response, and he actually does it. It is a flowchart for reliability: knowledge eventually leads to correct actions.

However, I would like to make three remarks. First, there are all kinds of different experts. There is an abundance of experts: for every small job there is an educational programme, a training course, a curriculum. You can become an expert in an endless number of fields, and in nearly every field. But there is no course in which you can learn how to do your humaning, and in fact this doesn't seem to be a specialism, doesn't seem to warrant a specific type of expertise. Society has no accredited institution that attends to the expertise in the field of humaning. Humaning is an activity that seems to be unrelated to expertise. After all, you can't leave your humaning to other people: everybody has to do his or her own humaning and we seem to need a whole lifetime to become even slightly competent at it. Eventually we may become an expert, but only at our own humaning, as there is nobody who can do our humaning for us. So even if the present-day construction with all its experts were a good solution for our problems regarding the reliability of other people, it still wouldn't work in the field of humaning. How well another person does his humaning is not a question of expertise, or of having a certificate with a stamp of approval. What matters, what is in fact the crux of the matter when it comes to humaning, is being reliable.

My second comment is related to this. There is something odd about the flowchart for reliability that I discussed above. A thermometer is reliable because it reacts systematically to temperature: its reliability is a function of the causal mechanism at the basis of its construction. The fact that it is 23 degrees Centigrade outside causes the thermometer to display '23 °C'. For experts, it doesn't really work like this. They take what you might call three-part action: first, a mechanic needs to know what is causing the noise, then he needs to know what to do, and finally he needs to actually start doing this. The first step is never simply a matter of causality for experts. And neither is the second, by the way. The systematic reliability of an expert is a function of his or her knowledge, and knowledge itself is not enough for the third step. We all know this. To take a rather cheap shot: a bank manager knows what the figures mean, and therefore knows what he should do, but still he gives himself, against his better judgement, another major bonus. This step from knowing to doing, this third step, is not a matter of expertise; rather, it is the step that is traditionally known as 'being of good will'. The expert may have a wealth of expert knowledge; still, if he is not of good will, he will not be reliable. If he doesn't know his job, if he is a cowboy, then he is not reliable either, of course. But however great an expert he is, that will not take him beyond the second step: knowing what needs to be done. Something else is needed for actually doing this: that requires a person of good will, a person who can do his humaning, a person who takes the quality of humaning seriously.

Lastly, my third comment turns that picture around and makes clear in one go that the so-called safe option is a dead-end street, and an incoherent undertaking. Let's have a look at the two possibilities that I presented at the beginning of this section: either you begin by entrusting somebody with something because you dare to carry the responsibility for that, or you begin by determining whether someone is reliable because you want to be certain that he or she can bear the responsibility involved. This second, so-called safe option requires that you start by determining someone's reliability. This entails that you yourself can shoulder the responsibility for determining this. And that responsibility never goes away. It is the flipside of the other person's good will that you always need. You can take a look at somebody's certificates, you can study the accreditation of the training course he followed, you can request the inspection reports of the review panel and work your way through their dry sentences, you can scrutinise the stamps of approval, you can examine the inspectors' CVs, and you can enquire at the Ministry. It doesn't matter how far you take your quest for solid ground, at some moment you will have to face your own responsibility and you will have to realise that you need to start having confidence, you need to start building on your judgement that the inspector is reliable, or the person who appointed him is, or the procedures that he follows are. You need to start taking responsibility, just like, long ago in your cot, you started entrusting, passionately and completely.

There is no way around it. If entrusting is an explicit, conscious activity, if you need reliable fellow human beings, you will need to start taking responsibility for your own vulnerability.

5. Vulnerability

The safe and smart option was described in the previous section: building on reliability, on knowledge, on expertise, on the joint escape from human vulnerability, on investing in the division of labour in society, on doing for other people what you are good at, on specialising, each in his own way; all in order to benefit from all our talents and to demonstrate that our society, and thus the way we live together, can be remodelled to suit our own perception of it.

This view is too simplistic, I know. This collective power, even if it were imaginable, realisable and achievable, can only exist thanks to our human trust, thanks to the fact that we sincerely dare to entrust other people to do their humaning in a reliable manner. The trust we give makes us vulnerable, each one of the seven billion human beings on earth. However, we seem to have forgotten that we need this trust; we have started to believe that we can insure ourselves against unreliability, that we can fight our own mistrust with expertise. Just shows you how naïve you can be.

In this respect it is hardly surprising that our society is such an emotional shambles nowadays, as all this sensible expertise is based on denying our vulnerability, on ignoring our emotions. Thus we miss the chance to enrich our emotions, to transform them into the important building blocks of our common sense.

To illustrate what I mean, I would like to have another look at Dick's anger and Thebe's sadness. Dick has quite a good reason for being angry: he thought that he could trust Ningak and Katya to be good at humaning, to deal respectfully with Thebe's love poem, just as is expected of good, decent people. Thebe had her own expectations: she expected Ningak to be worthy of her vulnerable intimacy, and that he wouldn't expose her. And Dick didn't expect Katya to lie or to meddle in the confidential correspondence between Ningak and Thebe; he expected her to respect their privacy. Even though Dick is not directly involved in this matter, he feels entitled to take Ningak and Katya to task for their behaviour. And he has a point: Ningak and Katya's humaning was definitely below par; it is appropriate to correct them. However, Dick's anger is awkward and confuses the other people present. Dick's anger makes his position less clear and thus also obscures Ningak and Thebe's position, as well as the position of you and Esteban. Consequently there is a lack of clarity, which results in a great deal of misunderstanding.

Dick's anger is reasonable to the extent that Dick is directly showing how he assesses the situation, thus giving himself the responsibility that should partly be shouldered by the other people. The quality of the moral climate on Endoxa is close to Dick's heart, but it is everybody's affair, not just his. This is why Dick feels left in the lurch, too heavily burdened, vulnerable in his feelings of responsibility, dependent on other people who seem to be unreliable in this respect. And this is where his anger takes over. Apparently, at this moment Dick is not capable of transforming his emotion. He chooses the wrong words, overstates his point, lets rip, and thus displays his vulnerability in a way that can have even more negative effects. Will the other people be capable of approaching Dick with compassion? Will they be able to respond adequately to Dick's assessment of the situation? Or will they, as a result of their own emotional vulnerability, react primarily to his angry words rather than the reasonable underlying meaning of these words? Thus you can see how misinterpretation can get out of hand, going from bad to worse.

Such a distressing ending would certainly not be surprising. People are vulnerable, dependent on other people's understanding, their ability to trust the other person's good will. Such a scenario will end in a regrettable manner if they cannot show this trust, if they are not capable of responding with trust, and if they are not able to recognise their own vulnerability. Such an ending would be understandable, but not efficient and definitely not constructive.

However, if someone is capable of utilising his own vulnerability at such a moment – whether it be Dick, Katya, Esteban, you or someone else – such a person will manage to really transform and enrich his emotions. And such a person would introduce a novel interpretation of the scenario, a new interpretation in which all the people involved are stimulated to enrich their emotions. And thus, it also becomes clear that trust is closely related to accommodation.

6. Accommodation

When Esteban took your hand, squeezed it lightly and then held it for a long time – that felt really good. This hand is the hand of accommodation, of compassion, of being allowed to be what you are emotionally, of the pioneer who stimulates understanding of things that are as yet insufficiently clear. Esteban's hand could have been the hand of a mother, a teacher or a guide. I will leave Esteban's hand in yours. Enjoy! Instead, I will turn my gaze towards Katya and try to imagine what she could have learned from Esteban's touch.

Suppose things actually happened in the way you imagined as you lay pondering them in bed. Ningak is standing there, holding his anonymous love poem. Katya enters. They like each other but haven't told each other yet. Both are blushing, and both are embarrassed. Suddenly Ningak blurts out: Katya, is this yours? Did you send me this poem?

What would Katya actually do at that moment? It is easy to see that this scenario can go in all kinds of different directions. There's a whole range of possible responses. This shows the light side of understandability, a narrative criterion. What Katya will do depends on her intentions, on the plan she has or that occurs to her, a plan that may give Katya a certain kind of potency. In the scenario that I pictured at the beginning of this chapter, Katya does not know how to deal with her feelings of jealousy and she comes up with a deceitful plan. But let's suppose instead that Katya has the tendency to do her humaning to the best of her ability, that she tries to be a good human being, a person of good will. Suppose that this is her plan and that her behaviour will be an efficient contribution to a positive moral climate on Endoxa. She just wants what is best for everybody. Suppose.

Of course, she likes Ningak, and she loves herself. But that is exactly why it is important that she can muster up accommodation. Here is what she might say.

Accommodation

What do you mean, Ningak? A love poem? How sweet. You don't know who sent it? No, I could've done it, and might have wanted to do it if I had thought of it myself, but no, I didn't write this for you. It must be from someone who doesn't dare to reveal who she is, but who cares a great deal about you. I can see why, you are a nice man, but no, it wasn't me. Do you have an idea who it might have been then?

Many things are happening at once in this response. To be exact, I can see four actions in one, four actions that are characteristic of an intervention by someone who can underpin her common sense with a powerful dose of accommodation. A person who understands that people need trust to be able to do their humaning, and that in order to be successful, trust requires accommodation.[3]

First, Katya identifies and confirms that Ningak is touched by something that is both valuable and vulnerable. Ningak goes red. He is self-conscious, not because he is ashamed, but because he doesn't know what to do with himself. He has lost his emotional balance and is looking for support. His emotions have taken over. He is not able to act decisively, and he feels exposed. Katya's response actually confirms his emotional impotence. She makes clear that she understands that he is moved, and that indeed he has received something of value. A similar response might be given by a mother if she hears that her daughter has not been invited to a party, even though her daughter's best friend *has* received an invitation. The mother will confirm her daughter's sadness, and will not deny that a party is indeed fun and that it hurts if you are not invited. But of course this is not the only thing a mother would say.

Second, Katya identifies and confirms with her response that she has also been touched. Ningak is not just put off balance by something that he thinks is important but in fact isn't worth all the trouble. No, Katya places herself in Ningak's position and makes clear how it would affect her were she to be touched by love. In fact, she goes beyond that and makes clear how much she likes Ningak herself, which is something that could easily put herself off balance. In the scenario that you are pondering at night, this seems to have actually been the case. In reality Katya was off balance, and this is why she wasn't accommodating, but let herself go in her jealousy and her small-minded self-deceit. But in the scenario *Accommodation*, Katya manages to be there for Ningak despite her own vulnerability. And that is what also happens to a mother trying to cheer up her daughter. It is sad if you have not been invited to a party; there is no denying this, and so the mother shouldn't deny it. On the other hand, it is not the end of the world. There is no need for her daughter to be overwhelmed by sadness. There will be different parties to which she will be invited, and then there will also be other children who do not get an invitation. That is the way it goes with birthday parties.

Third, Katya gives Ningak enough scope to do some good humaning. She leaves him enough room for his own reaction and doesn't take over his responsibility. However, what she does gives Ningak room to respond in his own way. In a way, she places his discomfort in a wider perspective. She does not leave him to his own devices, does not take away his responsibility, but provides him with a map of the area in which he could take a next, confident step. This is a typical aspect of accommodation. It is similar to what happens at a party: if you are giving a party, you briefly show your guests around at the venue, show them where they can find the drinks and nibbles, who is in charge of the music, and so on. This is again comparable to the situation with the mother and daughter. The mother is not going to compensate for missing the birthday party by buying a present for her daughter or smothering her daughter with attention. She places her daughter's sadness in the right perspective and thus offers her child the space to assess her sadness in her own way. There is a party and you are not there. That is all there is to it.

Fourth and finally, Katya assumes the responsibility for the framework in which Ningak can reclaim his emotional balance. Katya does not only give him the space to identify and acknowledge the meaning of his emotions, but also explicitly assumes responsibility for the context in which Ningak can recover as a human being, so that he can continue humaning. Katya would like Ningak to go on humaning, and sees that Ningak is uncertain about his own humaning. In the current scenario he is not sure how to proceed. Emotionally he puts himself in Katya's hands, hoping that she will actually do his humaning for him. He doesn't do this in any rational manner or with a clear understanding of Katya's expert reliability though. In a way he just throws himself desperately into her hands, comparable to the way the daughter throws herself crying into her mother's arms because she wasn't invited to the party. And then it is also comparable to my guest arriving at my party, who clearly doesn't know anybody and is standing there, doubtful whether he should try and make contact with the people on his right or on his left. Of course, we go up to such a person to help him on his way. And that is just what Katya does. Ningak has to determine his position. She's not going to take that out of his hands, but neither is she going to abandon him. She listens to what he has to say, offering him a safe environment in which he can think and determine what to do. Katya simply names things as they are, and as a result Ningak no longer has the opportunity to avoid his responsibility. Instead he has a framework in which he can simply trust Katya and is stimulated by her to trust himself too.[4]

What Katya is doing here is similar to what Esteban's hand has done to you. By taking your hand and not letting it go, by stroking your hand in a soft and comforting manner, Esteban has given you the opportunity to identify your feelings, to acknowledge these feelings, knowing that both of you have been touched by what is valuable to you both: your mutual proximity. Esteban has sensed this correctly, and in an accommodating manner he has dared to take the responsibility to make you feel that you can trust both him and yourself. You are doing your humaning together.

And it is easy to imagine that in a different scenario, Dick would have been better at dealing with his anger: he might have been able to transform it and to react with accommodation. In such a scenario, Dick could say something to Katya and Ningak that is comparable to the four things that Katya did in the scenario *Accommodation*:

Accommodating

I can definitely imagine that you don't know what to do with an anonymous love poem, Ningak. And Katya, I think everybody on Endoxa has noticed that you think Ningak is a great guy. He is, isn't he, Thebe? But still, pretending that you wrote Thebe's poem yourself, Katya, is not the way to deal with this. We have all seen the effect on Thebe, and her response is quite understandable, don't you think? Perhaps I shouldn't interfere and I'd rather not do this, but I would like all of us here on Endoxa to feel comfortable. And I think that means that everybody needs to show respect and understanding. Don't you agree?

7. Trust

We are all of us single embodied specimens of *Homo sapiens*. When I drink good espresso in the morning, only I can taste the delicious blend in my mouth. And only I notice that the caffeine does its work: I can feel myself becoming more alert. And when I'm having breakfast with my loved one, only I can taste the cereals that I am eating, and only she can taste her own toast. I am the only one who can remember my own dreams, and I am the only one who can feel the back rest of my chair against my back, who can feel my left foot against the table leg, who can feel the spoon that I am holding, who can feel my glasses pressing slightly behind my ear. And only I go to my own work, on my own bicycle, only I can feel the wind blowing through my hair, can think about my own plans and appointments for the day, can sit at my own desk and write these words. I'm always only myself, the one who always accompanies myself, in all my experiences, memories, plans, doubts, actions, emotions, considerations and so on. I exist because of my body; I *am* my body, and that is all. That is it, with a net weight of 80 kilos, with 200 bones, 600 muscles, more than 8 metres of intestines, 100,000 kilometres of blood vessels, tens of billions of nerve cells, all wrapped up in less than 2 square metres of skin, and this skin forms the limitation of the whole job lot, separating me definitively and forever from the whole universe. That is me. Isn't it?

No, not at all actually. Not only because I breathe, let air stream into and out of myself, and because I am part of my environment in a material sense through my digestive system. It is also specifically because I cannot do my humaning on my own, because I fundamentally cannot exist by myself: I cannot be what I am and do what I do without those seven billion other people joining in. Together we actualise my behaviour, our behaviour. If I do my humaning, others have to join in for it to actually *be* humaning. This started when I was born and will only stop when I die.

I need other people to make sure that my behaviour means what I expect it to mean. I need other people to understand my behaviour, to make it efficient and good. And I need other people to deal with my emotions, to discover what I care about, what is important to me. I need to entrust other people with acknowledging me, with doing their own humaning so that I can do my humaning, with being reliable, with being people of good will. I need to trust them so that I can be who I am.

It is when we are actually living together that our common sense develops. Our common sense becomes a wonderful, powerful force as we attune to each other; as we are interminably watching, refining, reassessing and correcting our expectations; as we are constantly managing, confirming and reinterpreting our more or less implicit ideas about the understandability, efficiency and goodness of behaviour; in our permanent, detailed and common balancing. Our natural power lies in our common sense, our ability to deal with our emotions, our sensitivity and our vulnerability. It is our common sense that knows how to deal with our passion, our passivity and our passionateness, exactly because our common sense thrives on trust and accommodation.

It is thanks to our common sense that our emotions transform, that they are enriched and become a reliable guiding star. In this way they become more than an accomplice and a messenger. They constitute our common sense. They are at the front when things are easy, raise the alarm when things get tough, and encourage us to assume an investigative attitude when we are surprised. In that grey area, where we have no choice but to be amazed, our common sense is in its element, and this is what I would like to show you in the next chapter.

Notes

1   Blackburn (1998) argues forcefully for the crucial role of our cultivated emotions in our commonsensical moral outlook.
2   My thinking on the role of trust in the development of our common sense is seriously influenced by some recent papers, notably Baier (1986), McGeer (2002), Hieronymi (2008).
3   I have argued more carefully and more substantially for this account of accommodation in Bransen (2014).
4   The framing of Ningak's emotions as provided for by Katya's accommodation very much looks like what Fonagy *et al.* call 'mentalization' in their publication of 2004.

5 Dealing with grey

1. A new job?

Thebe is giving you a long, hard look. She is inquisitive while she is talking to you, taking pauses of 2 to 3 seconds:
— The kitchen . . . The jetty and the harbour . . . And the Old Fort . . . The kitchen . . . Yes, and now the clean-up crew . . . The campfire.

She goes on for a few minutes. She looks at your face: in full concentration. What is it that she sees? What is it that she is looking at? Kondrat and Gabriella are silently observing what is going on. They seem fascinated.
— The kitchen . . . The Old Fort . . . The infirmary.

And then she suddenly says:
— Okay, it is clear. Quite clear. I feel no hesitation at all. You need to do it. Working in the kitchen is really the thing for you. That sparkle in your eye – to the right. And there on the left, next to your nose: such a clear *dinkla*. Absolutely clear, really! There is no room for doubt at all.

Kondrat turns and looks at Thebe. He is silent for a while, and then takes a deep breath.
— Well? Is she right? Are you going to do it?

There you are. Are you going to do it? Anushka has asked you if you would like to come and help her in the kitchen. That would mean you'd have to stop working in the harbour. Quite a change. How will you like that? You're so used to working on the quay. It is fun; you're really enjoying your time with Esteban there. But still, the kitchen. Cooking dinner. That sounds attractive too.

You have been thinking about it for a couple of days, but you're still not sure. You have told Thebe. Kondrat was there, and he was immediately excited when Thebe started telling all of you about what she did before she arrived at Endoxa. For decades she had been working with what she called *dinkliri*. Anyway, it was something like that. You're not quite sure whether you have remembered correctly. The word didn't mean anything to you. It did mean something to Kondrat though. At least, he said he had read about it. He explained the idea to you. It was a way, he said, to map the depths of a person's motivational profile. Something like that. Kondrat and Thebe then discussed it for a long time, and Thebe had said that she might be able to read your face. And this is exactly what she was just doing.

You haven't the faintest idea what a *dinkla* is. And should you leave the harbour and go to the kitchen? You haven't decided yet. You ponder.

2. Grey

Things are usually simple and completely clear. Your alarm goes. Time to get up, have a shower, brush your teeth, get dressed, and then have a cup of tea and piece of toast. If you enter the living room, yesterday's newspaper may still be there with a half-solved crossword puzzle, or the book that you were reading only last night. They may be there, but you won't really see them. At least, there is a great chance that you won't even see them, as you are now doing something else. You are in a hurry. The new day awaits you. You need to have breakfast.

It is quite simple, black-and-white you might call it. Breakfast: yes. Finish crossword or go on reading your book: no. Cup of tea: yes. Glass of beer: no. This clarity accompanies you all day. You will constantly find yourself in scenarios in which you know what to do. Somebody wants to get off board: your hand reaches out. The bell is ringing: go to the dining room. After dinner you stack the plates and give them to the washing-up crew. If it gets dark, you switch on the light. Simple. Just like old times at home with your parents. If a man asks you for the time, you look at your watch and answer him. If you hear a funny noise in your car, you go to the garage. If you go to a party, you bring a present. If you go to a restaurant, you ask for the menu and choose something nice. And afterwards you pay the bill. Tip? Ah . . . A grey area. You can make it simple for yourself. Simply never give a tip. Or always give 10 percent. But then when you're out with other people, the grey area merely returns. Because they do it differently. Or the food wasn't good at all, it was undercooked, or cold . . . Or the food was more than delicious and the service was really top notch, extraordinarily charming? Again . . . grey.

It is not just in scenarios in which it is *physically* our move – to put it like that – that there is usually the clarity of black and white. This clarity can also be found in scenarios in which our reaction is not so much a certain action but rather an attitude, an intention, a judgement, or a mood. It is also in such scenarios that this clarity of black and white is fully expected and usually goes without saying. On the beach there is a half-eaten hare. Its innards are lying around it in small pools of blood; strangely enough it looks as if its big brown eyes are begging you to help it. It is horrible, yuck! In another scenario, a baby's face suddenly breaks into a beautiful smile when you look at it, and you melt, of course. Someone asks you how you liked the cake that you are eating. Delicious, really! Katya asks you to go easy on the sugar as we are nearly out of sugar and the new sugar will not be delivered until tomorrow. No problem, of course. And so on.

But here there are also sometimes grey areas, a lack of clarity, an implicit difference of opinion perhaps, some confusion or underdetermination. And this is the grey I want to address in this chapter: the grey that makes us uncertain, that makes us hesitate, that makes us realise that we do not know, that makes us feel stuck, and that *makes us think* – and this is where in the end we find our greatness and our strength as human beings.

It is difficult to pinpoint what is so interesting about this grey area – or so complex or problematic. We cannot make this grey simple. We cannot pretend – certainly not – that it is simply black and white. We can't seem to stop making this grey difficult, you and I. And therefore this will be a difficult chapter: for me to write and for you to read. For you will not be able to simply accept that I have really caught in my description the essence of that grey that so confounds us.

Still, the beginning is simple. At least I think it is. Just have a look at the Animal Kingdom and see how its members seem to automatically know how to behave in their habitats.[1] It is as if they always know immediately what to do. For example, let's follow a ladybird in the grass. We can see how it slowly crawls to the end of a blade of grass, even though we have already seen that there is nothing there for it. Then we see how the ladybird only seems to notice this – if we can put it like that – when it reaches the end of the blade of grass. We then see it turn and go back. We can see it crawl from one place to another, aimlessly by the looks of it, continually busy, its reactions always triggered by the latest situation. At a certain moment it will open its shield, unfold its wings and fly away. What was the point of all that? No idea. It seems as if it doesn't matter to the ladybird. It just does what it does. It crawls where it crawls, flies when it flies, then starts crawling again and all the time we can simply conclude that it has no idea what it is doing, and that it is not thinking about what it is doing. It doesn't have to think about what it is doing because there's nothing to think about for a ladybird. It just does what it does. That is all.

Let's extrapolate this image of an uncomplicated ladybird just crawling around. It is the image of an insect for whom the triggers that it picks up from its environment are enough, triggers that are sufficient reason for it to give its follow-up reaction. Subsequently, it picks up the next trigger that occasions the following reaction, and so on. From beginning to end it is one continuous chain of action and reaction. There is never a moment of doubt. There is never a hitch in the chain, at least never a hitch which isn't also simply a link in the chain. Consider for example the incredibly intriguing displacement behaviour of the stickleback (Tinbergen and Van Iersel, 1947). Somewhere between attack and flight, between fury and fear, the stickleback bites in the sand. This apparently pointless behaviour seems to be invoked by a kind of internal short circuit, by two contradictory urges that temporarily are equally strong. Still, this displacement behaviour seems to cancel out the apparent ambivalence, so that the stickleback can go on. Or, to put it differently, the displacement behaviour of the stickleback cancels out the grey area that was there facing the stickleback. But sticklebacks can't work with grey. They don't know ambivalence; they wouldn't even recognise it. They cannot experience doubt, not in the way that characterises us as human beings, the way that encourages us to think, that invites us into philosophy's anteroom, as I called it in the introduction. Sticklebacks lack the equipment: their lives are all black and white. There is no grey. It's eat or be eaten, that's all. No grey. No questions.

So what do I mean by grey? Good question. Exactly, a *question*. A first question. A question that a stickleback never asks. Or, to be more precise: it is a question that will never be asked by a creature that cannot think, that cannot adopt an investigative attitude. And that is what this is about. Admittedly, it is only the beginning, the beginning that has been hiding between the lines of the previous three chapters. In Chapter 2 we discussed expectations, and folk physics and folk psychology. We discussed the normativity that is present in some psychologically based expectations, expectations that are related to entitlements and obligations. Sometimes some of these expectations are not met and then some room for disagreement opens up, room for you to think you are entitled to confront me about my obligations, which I can then counter because I meant something else than what you expected, or I myself expected something else than what you meant. This is the first kind of grey, the grey that Dick was confronted with in *Date*.

Chapter 3 was about our tendency to rationalise our behaviour, to account for it in an interpretative manner because we have no choice but to expect our behaviour to be understandable, efficient and good. There was a great deal of grey there too: grey that was hiding between the lines but that was sometimes unavoidable. Think about the newly married couple: the wife cooked more food every day, and as a result the husband felt he had to eat more food every day. Think about your own cooperation with Sybille, for example in *Apate*.

In Chapter 4 there was also a great deal of grey between the lines. Grey turns up anywhere where emotions have the upper hand, where people are passionate but also passive, or where people really go for something but at the same time are also being dragged along. And it is there that the grey turns up, because one person has a say in another person's actions, and because the meaning of someone's behaviour is also constituted by the response of the other person. Of course, at such a moment it is best to take a step back to see what somebody really means. This is what Dick could have done at the moment his anger got the better of him. But he didn't do this, and in this way he created a great deal of obscurity. It is also what Katya could have done when Ningak was so embarrassed when he started blushing and had no idea how to deal with it. But she didn't do this, and in this way she allowed a precarious scenario to get out of hand.

3. Questions

If you become aware of grey, you gain the opportunity to ask a question, a specific type of question, a question to help you understand. This is a question that ladybirds and sticklebacks never ask. It is a question that many people do not ask either, even though it is a question that is at the top of the list when we use our common sense.

I have to make a few distinctions in order to explain and justify this. First of all, there is the distinction between the question that you ask because you want something and the question that you ask to express your amazement. The first type is the question that simply asks for a certain type of response; this is why some animals seem able to ask such a question too: it revolves around getting something that someone else has and that you would like to have. Ducks do not ask each other this question. If a duck has got hold of a piece of bread, another duck can try to snatch it away, chase after it, start pecking at the other duck and so on. But looking at the other duck with a questioning look, in the way the ducks look at you when you are there with your child and a bag of stale bread, that is something that ducks don't do with each other. By the way, it is not really clear whether they are actually asking for bread if they look at you like this. They are probably just waiting for their chance to grab the bread. The fact that you are used to asking questions and giving answers may give you the impression that they are asking for bread. Interestingly, things are more complex among primates. Apes do not only try to take tasty-looking grapes from each other, they also seem to ask for them, as if they feel they have some sort of entitlement. This has to do with submission, with knowing your place in the hierarchy, as well as with the legitimacy of your needs; it is as if there are sometimes reasons for the more dominant apes to take the needs of the more submissive apes into account and to grant them their request.

For human beings this is an everyday occurrence. Questions are a part of everyday life that is completely familiar to us. If you do not know what time it is, you can ask somebody else. It is an easy question because the other person can tell you the time without losing it himself. It is more difficult to ask somebody on a bus if you can sit in their place, because that means that the other person has to give up their seat. Many questions of the first type, the questions you ask because you want something from another person, are therefore part of a complex economy, of giving and taking, of being more or less even, or of profit and loss if supply and demand are distributed unequally. In our current information society this has led to a complex knowledge-driven economy, and in fact this is odd because, like telling the time, you do not lose knowledge if you give it to somebody else. Knowledge can be freely disseminated, although there are some people who don't seem to like this because the production of knowledge can be so demanding and expensive.

The arrival of the knowledge-driven economy has led to certain questions, and these are questions of the second type, the ones you ask to express your amazement. However, there is a great chance that both you and the people who feel uncomfortable with such questions see them not so much as an expression of amazement but rather as an expression of criticism. This leads to the second distinction: between the question that you ask to encourage another person to think about what you consider is a good argument, and the question that you sincerely ask to express your amazement. Usually, the critical question is not an open question about the state of affairs that leads to amazement, but a question with an educational purpose, aimed at the other person's attitude and motivated by an idea of superiority. Such questions occur in all shapes and sizes, just like ideas of superiority and educational purposes come in different shapes and sizes. At one end of the scale there are questions that teachers ask to check whether their students have processed the learning material, such as 'What is the capital of Slovenia?' At the other end of the scale there are rhetorical questions, traditionally used to confront obvious lawbreakers, such as 'How do *you* feel about driving at 70 miles an hour through this sleepy little village?' These critical questions have in common that the person asking them does not seem to have any doubt about the correct answer. These are not the questions that are asked when you become aware of a grey area.

There is a third type of question that I would like to distinguish, namely the request for acknowledgement and recognition. This is a question that demonstrates common sense as a reaction to becoming aware of a grey area. While the critical question in fact demonstrates too much self-esteem, the request for acknowledgement or recognition actually demonstrates too little self-esteem. After all, it is about the confirmation of you being right, and eventually there will be no more grey but only black and white. Of course you know how things work, you know that you are right, but you cannot trust your own authority. You seek acknowledgement from another person. Asking yourself such a question would make it a rhetorical question. You know that you are right; you are looking for confirmation, not understanding.

This leads to the way in which a question asking for understanding can be asked in reaction to the awareness of a grey area. It is a question that you would initially ask *yourself*, rather than another person.[2] The grey is not grey because the other person doesn't understand you or you do not understand the other person, although mutual confusion may be an indication of the awareness of grey. But if you become aware of the grey as grey, it is because you realise that you are not sure what you should think about certain issues. What should Dick think about his own expectations in *Date*? What should Jayden think about his own behaviour in *Apate*? What should you think about Anushka's question about whether you would like to come and help her in the kitchen? And what should you think about Thebe's remark that this clear *dinkla* (clear *what*?) next to your nose makes it obvious that you will be in the right place in the kitchen?

We can find ourselves in an extremely interesting situation if we ask ourselves a question. It is a situation to which our awareness of grey can lead us. It is the situation in which the answer cannot come from another person, but will have to come from *you*. You will have to be able to understand. You *want* to understand it yourself. It is an intriguing situation because you are asking the question and must surely realise that the grey is grey rather than black or white. At the same time it is you who will have to provide the answer, who will have to turn the grey into black or white. Of course cheating will be of no use. Simply shouting something or simply giving a random answer is not going to convince you if you have asked your question sincerely. It is therefore a situation in which something really needs to happen, in which you have to go through a mental process, in which you will have to *think*, and will have to discover what you are really asking, what you really want.

4. Desire

The power of our common sense is clearly expressed exactly at the moment that you ask yourself a sincere question. It is exactly this question – what do I want? – that is the question of our common sense. Try and imagine a concrete situation. You are sitting at the table and Thebe has just concluded something impressive but also unfathomable about your state of mind and about what you should do. Kondrat is heightening the tension by looking at you enthusiastically and asking you excitedly what you are going to do now. And you? You do not know and ask yourself that question: shall I promise Anushka that I will come and assist her in the kitchen? Is this what I want?

This is not the question that would immediately come to mind for a theoretical discourse on the workings of our common sense. This is understandable. Both 'common' and 'sense' point outwards rather than inwards. Moreover, both words seem to be more related to answers than to questions.

'Sense' refers to the impressions that we receive from outside, impressions that force themselves upon us and that themselves make clear to us what they mean. 'Sense' is related to 'sensible', to a response that is both self-evident and wise. The combination with 'common' adds two extra connotations, the first of which can be found in Aristotle's work. For Aristotle, our common sense is our ability to connect the stimuli received through our five different senses: you do not only see a black shape move about, you do not only hear hooves on the ground, you do not only smell the warm sickly smell of manure, but you see and hear and smell the commonality in those different impressions: a horse. The other connotation adds a different dimension of commonality, so that the concept also obtains a normative dimension. It is especially Reid and Kant who have emphasised this aspect of our common sense. If you react with common sense, you react in the way people generally react if they rely on the humanity that they have in common. Common sense refers to our way of reacting, a way that is normal, not only because it is what we are used to but also because it is how we are supposed to react, as it is a correct, appropriate response.

But if this is the meaning of common sense, what has this got to do with desire, and with the question that you ask yourself when you become aware of a grey area: what do I want? To make this clear, it is important to emphasise that this chapter deals with a special type of scenario: scenarios in which you seem to get stuck, scenarios in which our common sense is working most meaningfully, even though these are scenarios that we do not automatically associate with common sense.

As I have shown in the previous three chapters, our common sense does its normal everyday work in our expectations (our folk physics and folk psychology), in our rationalisations that make our behaviour understandable, efficient and good, and in our trust and accommodation that help us enrich our emotions. In other words, in everyday scenarios we do what we do, and life runs its course because everybody uses their common sense. Because everybody reacts with trust and accommodation to each other's expectations, each and everyone of us can make something understandable, efficient and good of our lives.

Behind this normal common sense in everyday scenarios, there is our elementary desire for a good life. After all, this is what it is all about, for each of us, at each moment in our lives, on Endoxa and elsewhere, now and in the future as well as in the past. I admit it is a bit of a hollow phrase, especially in such abstract terms, but in all its simplicity it is also a truism.[3] Everything that lives wants to feel good, that is what life is all about. That's all. Or rather: it's all about a good life! For us human beings this means that we want to *do our humaning* – this is equally simple and equally abstract. Aristotle was simply right. Just like sticklebacks want to do their sticklebacking and ladybirds want to do their ladybirding, human beings want to do their humaning. We want to do it well, and feel good, and for us human beings that is more complex since we are *zoon logikon* and as such we are always busy trying to interpret our lives. We cannot simply let things *be*, but we have to actively interfere in how we *should do things*.

As long as this goes like a dream in everyday scenarios, nobody worries. That our common sense is busy in those everyday scenarios and is doing a great deal of work behind the scenes shouldn't worry anybody either. But what we should keep an eye on is that in all these everyday scenarios, our common sense is guiding the whole thing so that we are prepared at the moment of truth when we all of a sudden find ourselves in a scenario in which it counts, a scenario in which we become aware of grey and things don't seem to be running automatically. It is in such exceptional scenarios that our common sense needs to be the power that helps us further, that can nudge us up over the hill so that afterwards everything can run smoothly again.

We already met this type of scenario in the introduction when I discussed the investigative attitude. I can now give a more detailed analysis. You are there at the table. Thebe knows what you should do, and Kondrat is looking at you impatiently. It is clear what they expect of you and you can easily describe this in terms of obligations and entitlements. Of course Thebe is not entitled to force you or persuade you to take up kitchen work. But she does consider herself entitled to give you advice. You have given her that entitlement yourself by telling her about your doubts and by agreeing to participate in her investigation, at least if that is what she calls *dinkliris*. There are obligations involved in this entitlement, both for you and for Thebe. She will have to explain what the advice is based on, what her reasons are to advise you to start working in the kitchen. And you are obliged to listen to her. But of course you are also entitled to continue asking questions, to remain doubtful, to go against her advice, even if you think her arguments are convincing. If so, you will obviously have some explaining to do, especially as Thebe receives all kinds of entitlements the moment you are convinced by her arguments. After all, that is what is involved in our tendency to comment on our behaviour, rationalise our behaviour, interpret our behaviour as understandable, efficient and good.

There is more. I can imagine that Kondrat irritates you a little. He is too eager and too close. I am not sure about his trust, but he seems to be a little short on accommodation. He could give you a bit more room to come to your decision. And what about Thebe? Her score is below par for both trust and accommodation, if I may say so. Of course I am in cahoots with myself here: I have described Thebe as the stereotypical expert who knows everything and who feels at ease in her role as guide and key advisor. She seems to think that her methodology is more reliable than your ability to make a good assessment. And she makes optimal use of the accommodation that you offer her, although she doesn't have much accommodation for you. It is completely clear to her: you must do it. Even your hesitation conflicts with her expectations.

There is a clear opportunity to miss the grey area in a scenario like this, to see the situation in Kondrat's or Thebe's black-and-white terms. In a sense Kondrat doesn't allow you the time to ask the question that your common sense has put out for you. He can hardly deal with the deadlock; he doesn't want to remain in a vacuum and increases the pressure. Kondrat seems to be suggesting that the question that it is all about, *what you want*, can no longer be taken seriously. In this scenario, Kondrat represents the general public that pulls us along on the road of progress, holding us hostage with expectations that seem to be blindingly obvious. Especially when experts such as Thebe have shown their light on an issue, it is no longer easy to adopt an investigative attitude, to see more questions behind the facts. And Thebe has given you the facts. Even though on Endoxa her expertise is no more than curiosity because there aren't the societal arrangements that used to give her *dinkliris* a natural authority, her advice can still lay claim to a certain authority. How is that possible, when the question on the table is actually about *what you want*? What will happen to this question if other people appear with their expertise?

My diagnosis is about the difference between the missing information subroutine and the investigative attitude. As an expert, Thebe has operationalized her questions: she has made them answerable and has changed them into questions about information that is missing. That is part of her methodology. She is into *dinkliris*, whatever that is, and for her, the question whether a certain task is suitable for a certain person has become a question about the sparkle in an iris and the clarity of the *dinkla* next to your nose. However, your question is different. Your question is not whether the person who you are is suitable for the task in the kitchen. Nor is your question about whether or not you have a *dinkla* on your face. Your question is what you want. Simple, and vague at the same time. This is a question that can be asked in more than a thousand ways and that can also be answered in more than a thousand ways. Of course your question can be reformulated using Thebe's vocabulary, but behind such reformulation, behind such an operationalization, there lie a great number of questions, questions that do not surface on their own accord. Will you lose the meaning of your question if it is reformulated in terms of *dinkliris*? What good is an answer to you if it is about *dinklas*? How can you translate such answers to your own behaviour? How can you integrate such answers into what humaning means to you, into what makes you tick, into your desire for a good life? What is Thebe's role in your self-understanding? What does her expertise do to your life? And so on.

What is actually happening to your desires, to your elementary lust for life, to your aim for a good life, to your humaning, if you reformulate the question about what you want into a question about missing information? That is a long version of the question of your common sense: *What do I want?* It is a question, but not about missing information. It is the question that comes before missing information, in a logical sense – not necessarily in a chronological sense. It is a question that requires an investigative attitude. It is a question about understanding rather than a question about information; in fact, it is a question about what this missing information might mean. You can even put it as follows: the question *is the desire for understanding*. This question is what humaning is all about, what the verb humaning actually is. As a person, you are a humaning creature, a *zoon logikon*, a talking animal, an animal that wants to be at home in the language that gives him his self-understanding. Thebe's expertise may be useful and she may trigger your self-understanding, but only if you have enough common sense, since you need to be able to adopt an investigative attitude and pay attention to all the questions that lie hidden under her operationalization of your question.

5. Money

There is nothing intrinsically wrong with operationalizing the question "what do I want?" There is little to be gained by only sitting and staring hopelessly at the question. And it may also be wise to make the question more general, since we can learn a lot from other people's experiences. What did other people do in similar situations? How did they determine what they wanted most? Did they like it? Of course they are different from you and their situation will have been different from yours; still, there will also have been many similarities. Therefore, an obvious first operationalization may be reformulating the question in terms of 'someone like you'. What kind of work does someone like you enjoy?

Our common sense has no problem with such operationalization. However, it will have to maintain a sharp focus on what exactly the grey area is and why it is a grey area to you. After all, there are countless ways of perceiving 'someone like you': a Brit like you, a senior citizen like you, a 5ft 4in woman like you, a transsexual like you, a Chelsea fan like you and so on. Your common sense will advise you to adopt an investigative attitude to see whether the insistent operationalization of 'someone like you' will give more nuance to the relevant grey area. How would you describe yourself? How do other people describe you? Do their interpretations help you to distinguish black from white? What are Thebe's reasons to think that someone with a *dinkla* like you would be particularly suitable for work in the kitchen? What is the grey that her advice can solve, that her advice can separate into the black and white that it had originally consisted of?

Dealing with grey often requires additional knowledge, information that is missing, but at the same time also the ability to deal with that mediating knowledge. That is a double ability: on the one hand the ability to make the transition from desire to the operationalized question and on the other hand the complementary ability to go in the other direction: from the information found to the satisfaction of the desire. This ability deals with answering the questions that accompany the central question "what do I want?" The first question asks "what is the information that I need to know what I want?" and the second one asks "how can this information help me to do my humaning?" For example, you had no idea what kind of *dinkla* you have next to your nose. You didn't even know that you could have such things on your face. And now that she has told you this, you still don't know what to do with this knowledge. You can compare this to the fact that it has recently become known that some people have a genetic disposition to addiction: how will you do your humaning if you have just heard that you have an increased risk of addiction? Do you now feel the urge to ask the experts how to deal with this?

As far as I am concerned, this doubly functioning interpretative ability should be an integral part of our common sense. We should not be lulled into a false sense of security by the experts who give us their information but who do not think about the question whether we actually need this information or what we should do with the information.[4] In fact, the work that has to be done by our interpretative ability cannot be taken away from us, as it is a variant of what we do when we do our humaning: interpreting our actions and using language to shape these actions. However deeply experts may infiltrate into our daily lives, we will always have to deal with these two transitions ourselves. Confronted by grey areas, we will have to transform our doubts into an operationalized question, for example a question about missing information. Then, we will have to transform the information received from the experts back to our own situation, so that we can reinterpret the fuzzy grey that is bothering us into a clear black and white. Thebe may tell you whatever she likes, but in the end it is you who will have to say 'yes' or 'no' to that new job in the kitchen. Whichever way you look at it, if you want to follow Thebe's advice you will have to reinterpret your own wishes into a need for Thebe's information about your *dinkla*. And then you will have to interpret this information as a crucial argument for or against working in the kitchen. Even when you hide behind Thebe's authority, this is what *you* are doing, this is what *you* are then turning the verb humaning into.

Our common sense not only runs the risk of being excluded by our apparently natural need for information but also by our apparently natural desire for money. Have another look at yourself while you are sitting at the table, mulling over that job in the kitchen. Are you going to do it or not? What do you want? We have already seen one form that this question can take: a question about missing, apparently objective information about what it is like for someone like you to work in the kitchen. However, the question can also take a different form, in which it is about the contrast between the value that the kitchen job has for you and the value that the harbour job has for you. How happy will you be in the kitchen compared to your happiness in the harbour? If this is your question, a financial interpretation seems promising. Even though policymakers are nowadays exploring the possibility of the GNH (Gross National Happiness), we are still living in a world in which it is much easier to calculate the GNI (Gross National Income).

Of course, everybody will realise that you are distorting and indefensibly simplifying the question if you interpret the question 'Kitchen or harbour?' as a question about the money you would earn with both activities. But the power of money is exactly what makes it possible to compare two essentially different desires. It is quite an achievement. It is very easy if you can simply draw up a profit-and-loss account; if you can do this, if you can carry out the necessary interpretations, the mediating power of money will be enormous. And it brings great clarity: there will be no more grey areas anymore if you can define things using money. Two dollars is more than one dollar and less than three. It is as simple as that. It is black and white.

In our everyday lives, the profit-and-loss account is underpinned by our interpretative ability. What is the price of what you want? And how much are you willing to pay for it? If you have a good look at these questions, you can see a great deal of grey. Because what you want depends on what it costs: you do not want to have it at any price. And what you want to pay also depends on what it costs, and here too: not at any price. If you look at these questions using your common sense, many things are unclear, doubtful and underdetermined: there is a great deal in these questions that makes us think and that triggers our investigative attitude. But the strange and estranging thing about money is that an easy reversal is lying in wait. If I give you the choice between two identical products, one costing $100 and the other one costing $200, it is immediately clear which of the two you will choose: the cheapest. And if you can choose between two identical tasks, one that will earn you $100 and one that will earn you $200, it would also be immediately clear: you will choose the one that would make you the most money. In other words, in terms of money you always know what you want: more. Never less, always more.

The mathematical clarity of money seems to offer us a chance never to be bothered by grey again when we ask ourselves what it is that we want: we want more. That is simple enough. However, we then have to stay within the limits of the profit-and-loss account. Within these limits you do not need an investigative attitude anymore; your mathematical skills are enough. If you become ensconced enough in the banking world, then there is a chance that you will lose sight of these limitations, and it will become only about the money that you can use to make even more money.

In the context of this book, you will have undoubtedly understood that I am not unreservedly positive about this possibility. If you are getting used to thinking in terms of money, there is a strong possibility that you will overlook all the grey areas that are in front of you. If you are getting used to thinking in terms of money, there is a strong possibility that you will lose the ability to ask questions and that you become estranged from your own desires. Thinking in terms of money is ideal for accountants with well-balanced chequebooks, but it is an ideal that imposes great demands on our common sense. Merely thinking about the chequebook sidelines your common sense, the common sense that advises you to adopt an investigative attitude if you need to interpret your desires in terms of money, and your money in terms of desires. This common sense encourages you to look behind every dollar for the grey area that it is covering. This common sense knows that grey and money do not go together well. It is wary of easily waving away difficult questions and of the genial smile that people use when they admit that money may not be able to make you happier, but that "it surely helps!" This common sense knows that you cannot express lack of understanding in money.

6. Understanding

Time to recapitulate. I have been taking you on a long journey and I can imagine that by now you have the idea that you are lost, and you may be wondering about where exactly you have ended up, and I can also imagine that you have lost sight of the grey area that I was trying to make you deal with earlier. I may have warned you about this, about the grey that can be so problematic, about the problems of a mutual lack of understanding, but I am not so sure whether that is a good excuse. What on earth has this chapter been about?

Human beings do their humaning. I started my exploration of life on Endoxa in Chapter 1 with this observation, my exploration of the lives of everyday people who need to get by without expertise. I have argued that these people are fine using their common sense, and are able to adapt adequately to new situations. Their common sense helps them to recognise new, strange, unknown things as familiar things. I have developed this ability in Chapters 2, 3 and 4 as the ability that presents itself in the entitlements and obligations that accompany our expectations, in our interpretations of our behaviour as understandable, efficient and good, and in our trust and our accommodation that allows us to adapt to new situations in an emotionally adequate manner.

In all these different ways, our common sense is our ability to deal with potentially grey areas, with the white noise that is between us, with the differences in our habits, our tendencies, our desires: in the differences in how we do our humaning. Our common sense helps us to make a clear distinction between the black and white in all kinds of different scenarios. Our common sense helps us see in a sophisticated and detailed manner what the case is: that we are at home in our lives. We are at home in our lives, but in a different way than ladybirds and sticklebacks. We are not simply identical with what we do, but we can and do reflect on our behaviour. We can consider our behaviour from an external perspective, we can embrace it in an exceptionally intimate manner, and we can appropriate it as behaviour that shows us exactly how we are, and how we do our humaning. It is especially in this reflective attitude that our common sense is at its best. At least, that is what I claim.

This is why I have been exploring scenarios in this chapter in which there is a more persistent grey area, a grey area that doesn't only look grey because we are not looking accurately enough but that really *is* grey: doubt, confusion and lack of understanding. Because in the end, only when we overcome the real grey can our common sense show us what it actually is: common sense is our ability to deal with grey. In the first part of this chapter I localised this grey in your doubts about what you want: kitchen or harbour? I hope that this grey is recognisable and corresponds to a topical theme in these times of authenticity.

That may well be what is bothering you. Real grey is not an internal, intrapersonal affair, you may well think. Real grey is found *between* people. Real grey is *interpersonal*. Real grey arises because people have different ideas about how they should be humaning. Real grey can be found in our differences of opinion, in our political struggles or our mutual incapacity to understand the other person's motives or the other person's audacity to assert what he asserts. The real grey is in our intrinsic moral differences, in our incomparable views of life. You may think it's a shortcoming that I am avoiding the real, serious issues and that I am instead dealing with a contemporary theme that is too fashionable and trivial.

I can understand this reproach. But here we are discussing a potential grey area, of two different intrapersonal affairs. On the one hand, your doubts about the kitchen job indeed seem a matter of authenticity, of what you want most, of what gives you the most personal satisfaction. But on the other hand, this is an example of a situation in which you, as a human being, are really asking yourself a question. That is exceptional: asking yourself a question. Really very special, for if you do not know the answer, there seems to be no point in asking yourself. After all, if you knew the answer, you wouldn't ask the question. And if you do not know it, there is at least one person that you should not ask this question: yourself. Still, you can pose the question *what do I want?* only to yourself. And everybody understands in the kitchen-or-harbour scenario that this is exactly the question that you should be asking, that you are asking yourself that question and that this makes sense. An escape to the expertise of other people or to the clarity of a profit-and-loss account is exactly that: an escape. Trying to dodge the question. And that is what makes this a good example of what is going on when there is a real grey area.

This is also true in interpersonal matters, in serious moral issues. For example, it would also be true if you were to hear me state that the only workable didactical concept is based on randomly appointing a scapegoat in your class, so that the other pupils can become a close-knit group based on rejection – to introduce a painful and strange subject. My point is the following. I expect that you would not agree with me if I were to seriously defend this. You would probably think that I am wrong, and that I would have to be wrong. For you, there is no grey area at all: it is simply black and white. There may well be a difference of opinion, but there is no grey area, as I am simply wrong. If you want to discover a grey area here, you may need to ask yourself the question how to deal with me, given that I am clearly talking nonsense. Did you notice? You are asking yourself this question.

You may also want to raise the issue yourself. But if you do this, if you do this while paying attention to the grey area, you will have to start imagining why I would defend this backward didactical concept. For starters, you would have to stop calling it 'backward'. You would have to take an interest in my reasoning, and discover its rationality. You would have to deal with me using your trust and accommodation. You would have to respect my entitlement to regard this didactical concept as understandable, efficient and good. You will have to respect your own obligation to deal with me as someone who knows what he is talking about. If you can do all that, if you can manage well enough to see the grey that symbolises our *mutual lack of understanding*, you will notice that it has become your own intrapersonal grey. If you really understand the rationality of my favourite didactical concept, you have indeed discovered a grey area, you have taken it on board, and now you are – *intrapersonally* – in a fix because you consider a didactical concept simultaneously both backward and reasonable.

I can imagine that it is difficult to muster up so much empathy. I am simply wrong, you may still be thinking. However much you try to discover the grey area, all you see is black and white. You may well decide that there is no grey at all regarding this didactical concept. It is simply flawed, immoral, untenable, inhuman, or however else you would like to disqualify it. This conclusion brings you back to the other grey area that is really grey: the grey that comes dominantly to the forefront in serious issues regarding a mutual lack of understanding. The grey that raises the question: how will you deal with me, given that I seriously want to introduce an inhuman didactical concept in our education? There are two things that should immediately strike you: (1) this is a question that you are asking yourself, and (2) it is a variation of the fundamental question: what do I want?

It is in sincerely asking yourself the question what you want, that your common sense shows itself to be a response to the grey that you have observed. This question makes you adopt the investigative attitude and shows you that you are not ready to operationalize the question. You will realise that it is not about a lack of information but about determining what you wish. *What is it that you want? What is it that I want?*

Still, this is more than just an intrapersonal quest, since it is not only about *your* common sense but also about *our* common sense. In the case of my inexplicable defence of an inhuman didactical concept, it is not only about how you deal with me, but also about how I deal with you; in other words, it is about how we *deal with each other*.[5] This is where the question probably becomes very difficult, but my argumentation very easy, theoretically, as a plea for our common sense. After all, we have seen the argument, in the previous chapters.

We sincerely ask ourselves the question *what is it that we want?* because we do not simply want something from the other person. We don't want any money or information: we want understanding. I do not want to simply agree with you and I do not want you to simply agree with me. That is not the point. It is about understanding. That is why our common sense advises us to adopt an investigative attitude, to trust in the other person and to offer that person our accommodation. This is something that we can expect from each other, since we all have the entitlement and the obligation to be humaning, because we are talking animals, *zoon logikon*, creatures that together will continue to try to put into words what makes our lives, our humaning, understandable, efficient and good, however much grey we encounter or produce in our lives. It is undoubtedly an arduous undertaking, a continuous practice, without end. But theoretically it is clear: it is a question of common sense.

Notes

1   See Maturana and Varela (1998) for an interesting view on the biology of knowledge.
2   These are the kind of questions that play a major role in Harry Frankfurt's work on the rationality and love that is characteristic of human agency. See e.g. Frankfurt (2006).
3   The theory of humaning that I am sketching in this first part of this book is my casual attempt to make sense of this truism. I have been influenced and impressed by a couple of recent attempts to address this truism in a much more sophisticated and systematic manner: Haybron (2008), Tiberius (2008), Velleman (2009).
4   Just to be clear: I am primarily making a case for the investigative attitude as the essential characteristic of common sense here; I am not making impetuous accusations against dysfunctional experts. Experts are not dysfunctional at all;

usually they do excellent work. That is not the point. The point is what we do to ourselves if we allow experts to do their work well.

5   My argument and my conceptual resources differ from John Dewey's, but basically my view of our quest for common sense in dealing with each other is quite similar to Dewey's view of the intelligent and social conduct characteristic to human nature. See Dewey (1922).

Part 2

Living with expertise

6 Humaning today

1. Back and forth

It's time to leave Endoxa.

How surprised and relieved you all were when Brenda came to pick you up by ferry. The time on Endoxa had become a pleasant holiday, one of those holidays in which you look forward to being back home in your own comfortable surroundings.

How small your house suddenly seems! Such a narrow hallway and such steep stairs. Such a tanned face — *really deeply tanned!* — in the mirror in the upstairs bathroom. Such a soft bed and such a curiously familiar crack in the ceiling of your bedroom, grinning at you while you are lying on your back, contemplating that intriguing thought experiment on Endoxa.

Who'd have thought that common sense could be so progressive. I'd certainly never imagined it.

2. Self-evident

Human existence takes place against a background of what is familiar, what is thought to be true, what goes without saying and what doesn't raise any questions. It is against this background that we understand new situations (or don't understand them). It is against this background that strange, unknown situations are strange and unknown and make us think. It is against this background that we can explore and question strange, unknown, new situations, and as a result, may eventually understand them. Without this background, without this nest built in nature (as the influential German philosopher Arnold Gehlen[1] calls our culture), we would be hopelessly lost. It is because we are socialised, because we have grown up in a society with set routines and customs, that we as human beings can do our humaning.

Within this conceptual framework, it is natural to regard common sense as a conservative force. In Part I of this book, I argued that it is our common sense that gives us the ability to adapt to new scenarios. However, Arnold Gehlen would argue that this is especially the case because we are socialised: this has made us so adept at recognising and holding on to the familiar in what appears to be new and strange. In other words, my common sense is active when I do not deviate, when I respectfully adapt to the world order and the way of living that my culture has in store for me. My common sense constantly tells me to behave like a normal, English person. This is why I have a cup of tea when I'm upset, why I like walking in the countryside, why I watch the *Six O'Clock News* to keep up with what is going on in the world, why I complain about the shambles the English National football team are making of it, why I am proud of my country but wary of Europe, why I use a lot of irony in my conversation, and why I say 'sorry' all the time.

In that light, my return to common sense seems like a dubious, conservative undertaking. I may need to dare to learn some lessons from the intriguing fact that nowadays we are no longer so sure about things. For example, we are no longer sure what to think of development aid, whether to be enthusiastic about European unity or not, what is happening to our pensions, what we should do to have a sustainable world. Aren't these all indications that whatever is going to show us the way ahead, it is absolutely *not* going to be common sense? Shouldn't I be pointing the reader towards the two alternatives that are clearly evolving from this line of thinking?

As a first alternative, we may simply need firm, decisive politicians who bluntly say what is what and are not afraid to take tough, unpopular measures. If we want a future for our country, little can be expected from the familiar commonplaces that seem to keep the Western world in their grip. The multicultural experiment has failed, and there seems to be no place for our tolerant approach in the current harder, business-like world. Such common sense is not going to help us make it in this tough global economy in which China, India and Brazil are doing their utmost to compete us into oblivion.

Second, we may simply need a great deal of expertise, careful scientific research, randomised controlled trials and evidence-based interventions. We will certainly need scientifically trained experts who can save us from the pitfalls of our common sense. Our naïve ideas and our trust in the rationality of our cognitive skills may have helped us progress for a while, but by now Daniel Kahnemann has clearly exposed our way of thinking as fallible (Kahnemann, 2011). The world is simply too complex for our common sense; most of what our common sense does is preserve our prejudices.

Such reactions are understandable. They confirm the persistent tendency to think that 'having common sense' and 'being socialised' are one and the same thing. Such reactions confirm how manifold the obvious is, and how difficult it can be to notice that the obvious is not so obvious at all. Ironically, these reactions confirm how socialised we are, to what extent we are children of our time, and to what extent we tend to think *with* but not *about* our conceptual framework.

3. Dialectics

Human beings are talking animals. We are not only our body but also our voice. We tell the story of our lives in two different ways, both in our body language and in our common language. As a result, something interesting is going on in our lives, because it is impossible for us to hold the following two things simultaneously as true:

- There is only one story of our life that can be told in two different ways.
- There are two different stories of our life that cannot be reduced to one single story.

Besides, people do not only talk; they also listen. People ask for and give each other their reasons for what they do. As a result, our story is twofold in a different manner: not only as the story told, but also as the story heard. Therefore, there are two more things that are impossible for us to hold as true at the same time:

- There is only one story of our life that is told and heard by different people in different ways.
- There are as many different stories of our life as there are people, and these stories cannot be reduced to one single story.

The fact that we are animals that have language at their disposal has led to these duplications. It is our fate, as human beings, both in a positive and in a negative way. Language does not only make understanding possible, but also makes misunderstanding possible.

These duplications both enable and necessitate the continuous switching of the obvious to the problematic and vice versa. Of course it is possible to linger in the obvious; often this is useful and completely logical behaviour. If someone asks you what the time is, you look at your watch and answer this person. If a traffic light is red, you stop. When you are getting ready for school in the morning, you don't notice the remote control of your game computer lying on the settee, ready for action. And if someone tells you that English people cannot be trusted, you ask him what he means and why he is so sure. Do you see what is happening? How easy it is for you to switch from autopilot to the investigative attitude, and how obvious it is for you to sometimes regard other people's reactions as problematic? Clearly, this is what happens all day. One word is often enough for you to understand what is going on, but sometimes this is not the case at all. Either way, you know what to do: you switch from the obvious to the problematic, and back. Hey, what is that at the bottom of your bag? Let's have a good look. Ah, that is the tissue that you wrapped around an apple core yesterday. Someone starts to tell you a long story in the pub. You start by paying attention, asking questions for clarification, until you realise that this person has started halfway through his story and has clearly had too much to drink. An author argues in favour of a problematic interpretation of the strength of common sense. What? Let's read it again carefully. Ah, now I see what he means!

Our shared background is our strength, but also our weakness. Our shared background makes us see things that are not obvious and it is this shared background that tells us what things we can regard as familiar, unproblematic, about which we do not need to ask questions, which we do not need to explain further, because these things go without saying. However, as always strength and weakness are two sides of the same coin. Because what goes without saying *never really* goes without saying; this idea is simply built into the duplication that follows from the fact that we are *zoon logikon*. There is never only one story. Every story is both told and heard. Every story is duplicated, from itself, simply by being a story, by being part of language, by being the medium in which human existence is realised, the medium that human beings use to do their humaning.

This challenging and fascinating aspect of our existence can be found in the endless dialectic that our talking existence entails. You could problematise this, just as many generations of philosophers before you have done. You could try and carefully map this. You could try and disclose the logic in it. I am not going to be so bold. I would rather consider it in a naïve, informal manner, in a similar way to how Henri Rousseau looked at the jungle. And then I can conclude the following:

- Our common sense enables us to see the familiar in the unknown, and this can also be the other way around. If the familiar and the unknown can hide inside each other, our common sense can also help us to see the unknown in the familiar.
- If our common sense can switch in such a way, from familiar to unknown and back again, from obvious to problematic and back

again, timing, attention and moderation become important. When is it necessary to switch? When do you prefer the obvious? When is it necessary to emphasise the problematic?

- It also shows common sense to switch at the right moment.
- There is a strong temptation to look for the rules that should be the foundation of this common sense. There is also a strong temptation to design a method that tells you when to be critical and when to conform.
- In our current society we have institutionalised critical thinking. 'You'd better leave that to science.' The scientific method is the method *par excellence* that tells us when to be critical and when to conform, when to ask more questions and when to decide that something is evident.
- The scientific method is a systematic improvement of what is available in our common sense in rudimentary form. This method is wary of the pitfalls of our fallible daily thinking, and soundly based on controlled experiments that can be replicated.

Unfortunately for science, this would be a good moment for me to make a cynical remark about the myth of science's ability to self-correct (Stroebe *et al.*, 2012). That would be a bit mean, and I feel too much sympathy for the sincere exertions of all those scientists that do and have done their utmost to help us in our humaning. Besides, I see the dialectic in a different field.

Science has been built on the idea that it is, in principle, possible to tell the story of the universe. The language to be used for that should be a language that is radically independent of the context in which that language is used, and radically independent of speaker and listener. Obviously, the story of the universe cannot be a human issue, cannot simply be a question of people understanding one another or not. It is impossible that the story of the universe is anything else than a completely objective fact, anything else than a completely transparent reflection of how the universe has evolved from the Big Bang to . . . to what, actually? Theoretical physicists ponder these things intensely. Their string theory is way beyond me (and not only their string theory actually). I am really unable to grasp this theory and therefore I humbly take off my hat to these physicists.

But what I do seriously wonder about is why the story about us, about *zoon logikon*, about *Homo sapiens*, should be part of the story of the universe. Why should the story that people tell about themselves, the story in which they manifest their humaning, in which they try to determine what they want and what they have to say — why should that story be told in a vocabulary that is radically independent of the context in which it is used, and radically independent of speaker and listener? Why should we, *you* and I, try to understand each other in a vocabulary in which there is no trail left of our desires, emotions, expectations, of what is understandable, efficient and good? Why should we have so little faith in each other, in our communicative interaction? Why should we expect accommodation from the universe if we cannot muster up this accommodation ourselves because we have had to disregard our own involvement in order to practise science?

It is time for Ian Hacking.

4. Let's reclaim our common sense!

Since 1965, Ian Hacking has manifested himself in an impressive series of publications as an indefatigable creative and discerning philosopher of science. In all his publications, Hacking has been searching for the point at which the story of the universe appears to be a story of science, and at which the story of science turns out to be a story of human beings. Again and again Hacking shows how the story of people runs riot with us, with us people. How this story confronts us with ourselves, how it refers back to us and shapes our self-understanding in the vocabulary that is available to us in this story. Hacking has written books on a variety of subjects, but whether they are about the logic of statistical inference, about reasoning with probabilities, about experimental physics, about psychopathological diagnostics, or about the history of mental illness, he always focuses on what he calls dynamic nominalism.

Citations and scholarly discussions are not really suitable in a book like this, but I am going to make an exception at this point, not only because Hacking formulates his point in a particularly wonderful and concise manner, but mainly because I want to emphasise the affinity between his argumentation and my argumentation, and because I would like to pay homage to his influence on my way of thinking:

> The claim of dynamic nominalism is not that there was a kind of person who came increasingly to be recognized by bureaucrats or by students of human nature, but rather that a kind of person came into being at the same time as the kind itself was being invented
>
> (Hacking, 2002, p. 106).

In the context of this book, the claim of dynamic nominalism points towards the fact that in a world in which people are seen as objects that can be studied by behavioural science, a certain kind of person comes into being who behaves like an object studied in behavioural science.

It is not that such people have always been there, ready to be recognised as objects to be studied by behavioural scientists and policymakers; such a person came into existence at the moment that behavioural science invented him. Such a person mistrusts his own common sense, and thinks that in order to gain self-knowledge, it is necessary to have a detached, impartial, theoretical perspective. Such a person sees himself as an object, a complex organism, a plaything of genes, hormones, glucose and neurones. Such a person knows that the expertise necessary to determine what makes people tick is complicated and advanced. One day – which is both a good and a bad day – such a person will realise that he understands only very little about the concepts that he needs in order to understand himself. At that moment – and I hope it is *now* – such a person will realise that it is time to see the strange and unknown aspects in everything that has become obvious to him. At that moment, such a person will realise that there is no way he can avoid his *humaning*, that he is a *zoon logikon*, a talking animal that needs to do more than just listen to the experts who tell him incomprehensible stories about himself. Such a person realises that he has to talk, that he will automatically start asking questions if he no longer understands. Such a person realises that his questions are not about missing information, but about the fundamental questions that precede these questions, questions that express his wonder, questions that people ask when they adopt an investigative attitude.

These are the questions that I love, the questions that I hear buzzing at the gates of philosophy. They are philosophical questions, referring to the concepts that we have started to use in a casual and obvious manner. Even though gradually these concepts have become more and more technical, they have become concepts guarded by experts, concepts that you can only use correctly if you have the relevant expertise. These are the questions that you ask with your common sense, lateral questions that you ask because you cannot follow the transition between on the one hand your desire for a good life and on the other hand the exceptionally specialist and probabilistic expertise that experts gain on genes, hormones and neurones. In an accessible, well-organised and surprisingly simple text, Hacking describes ten different engines people use to acquire expertise in the behavioural sciences (Hacking, 2006). The first seven engines are aimed at discovering the facts and are related to the scientific method that we know from physics. The eighth engine is aimed at regulating the practice of human behaviour, and the ninth focuses on the role policymakers play. I would like to focus here on the tenth engine, the one Hacking calls 'Reclaiming our identity', and that revolves around resistance to experts, to the people who know and who will guard their expertise like jealous sentries. It is this tenth engine that has led to this book.

In the following four independent chapters, I survey a number of aspects of our lives in the age of the advancing behavioural sciences. I show you four times that there are good reasons to claim room for an ability that cannot be taken away from us, but that is under threat due to this advance of expertise: our common sense. In this second part of the book, I have again chosen to incorporate my argumentation in fictional stories, but this time there is no overarching story about an isolated world. In this part I have chosen separate sketches of everyday scenarios in which our common sense seems to have lost its way, intimidated by the dominance of a scientifically sound approach. It is the toolbox of modern science that is handed to us without thinking, as if we, whatever we are going to do, will want to do this in a way driven by scientific knowledge. It is like the TV commercial in which somebody lays the table, puts the mayonnaise and ketchup on the table, and only then asks the question: 'What's for dinner?'

But sometimes all this scientific expertise is not helpful at all. Sometimes the scientific approach is totally unsuitable for the problem that you are struggling with, like a crowbar that is of no use if a toddler pushes a hairclip into a door lock. In the following chapters, I will discuss four presuppositions about our modern scientific orientation:

- Science is our best idea about the growth of knowledge, and it is an idea that still evokes an association with an objectively existing reality that is being mapped in an increasingly good, accurate and complete manner.
- Our language is a neutral and potentially transparent instrument that we can use objectively to describe this reality adequately and accurately.
- We can improve the quality of our lives by working towards the improvement of our circumstances in a scientifically sound manner.
- Humaning is a question of factual behaviour that can be studied from a scientific point of view.

I will argue that these four presuppositions are disputable; furthermore, I will make the case that these presuppositions are indefensible. However, such strong conclusions are not really what I'm after: my objective is to shake you up. I want to distress you, to make you think, and to wake you from your slumber and to make you realise that it doesn't show common sense to tacitly assume that the scientific approach always leads to good results. In other words, science is not an instrument that you should always want to use. Just like a camera: it is often more important to take a good look yourself, instead of immediately taking a picture for later.

Note

1 Gehlen (1940). Even though Gehlen was a shrewd thinker, I always approach him with certain misgivings, because several times during the heyday of his career he was appointed to a chair that had been occupied by a professor who had been sacked by the Nazis.

7   Waking up without science

1. Universal prevention

She is a firm believer. The principal has just introduced her and she is on the stage, full of apprehension; the projector on the little table next to her is showing the first slide on the white screen behind her. There are about forty to fifty teachers in the auditorium, all listening attentively. It is a large school, and many of the teachers are interested in her lecture on the prevention programme developed by her and her colleagues at the Obesitas Research Institute. She is hoping to test the programme at this school. She has been visiting schools for weeks to explain what the institute can offer and what they want to research.

She starts with the facts, which are shocking.

'In the past 20 years the number of overweight children has increased from nearly 4 percent to over 17 percent, which is an increase of more than 400 percent. Fortunately, the number of children with obesity is much smaller; still, the growth rate is similar: 400 percent!'

She is an articulate speaker with a smooth, pleasant voice. She knows what she is talking about. She begins by explaining the prevention pyramid, and starts at the top. This may be somewhat unorthodox, but in the past few months she has noticed that it is easier for the audience if you start with the worst cases. Going down the pyramid, it is easier for the audience to grasp what prevention really is, and how important it is. She has come up with a good example, a wonderful picture of 'Uncle Bob'. At least, that's what she calls him. The picture is from an Australian website, far away enough to use in her presentation here in Europe. 'Uncle Bob' weighs 350 pounds. He sits lopsided in a chair, on one buttock actually, as there is no more room. You can almost hear him huff and puff; it is really a great picture: so striking, so poignant. On the table next to him there is a bottle of beer, a plate full of roasted meat and big dollops of sauces in different colours. Clearly, he's at a barbecue. Uncle Bob has a heart condition and his cholesterol is too high. On top of all that, he is a smoker: look, you can see the ashtray behind the plate, full of butt ends. There is some smoke spiralling upwards. One by one, she goes over the details in the picture, slowly building the tension.

Then she introduces the concept *healthcare-related prevention*. The teachers start shifting in their seats. You can see them thinking: 'Prevention? It's a bit late for that!' That is exactly what she's about to tell them.

'*Healthcare-related prevention* is still a type of prevention, because it is designed to prevent worse. It can be used to show patients like Uncle Bob how to deal with their condition. It shows them the possibilities that are still present: how they can integrate their condition and the associated limitations into a lifestyle that is as healthy as possible.' With some understatement she remarks that in the case of Uncle Bob there is quite some work to be done.

'But that is exactly why there are other forms of prevention', she goes on to say. The next slide shows a single layer under the bright red triangle, and this layer has a slightly less intrusive colour: orange. She continues her lecture by saying that healthcare-related prevention is the last resort for patients in which *indicated prevention* has not been successful enough.

'*Indicated prevention* is also aimed at individuals, namely individuals who are running an obvious risk, such as Uncle Bob's children and grand-children. But we could also be looking at a clearly overweight person who is visiting his doctor with back problems. You could simply prescribe physiotherapy and leave it at that, and this is exactly what doctors used to do. But it is much more sensible to refer such a person to a dietician, who can inform him about a healthy eating pattern. The dietician can draw up a programme and see to it that it is followed. This is what we call *indicated prevention*. There is no obesity yet, but there is a clear risk of obesity. And that is what indicated prevention is aimed at: the timely prevention of obesity.'

There are two additional layers under indicated prevention. She shows the next slide with the penultimate layer in soft yellow.

'This type of prevention is called *selective prevention*', she tells her audience. 'You can imagine how it works here at school: there are chubby children and there are slim children, children who play a lot of sport and children who do not play any sport, children who eat healthy food and children who snack a lot. It is not difficult to identify the children with an increased risk of becoming overweight and developing obesity. You could put these children together and offer them a special pro-gramme, and this is what is called selected prevention.'

She waits for a moment, letting the silence hang in the air. As so often at this point, she sees her audience feeling slightly uncomfortable. And then she shows her next slide. The bottom layer has appeared, and to a pleasant fresh and green background it says *universal prevention*.

'Of course', she says, 'it is not very pleasant if you take "the fat child" out of class. This is something you simply can't do at this age. This is why we have developed a prevention programme at the most basic level. It's good for everybody to be informed about the risks of being overweight, even slim children. It is good for every child to learn the difference between healthy food and unhealthy food, even for the children who bring fruit to school every day. Just as it is crucial that all children are informed about the importance of exercise. And even for sporty children it is good to do some extra exercise. By offering a prevention programme for all, you can achieve many good things in the children who need it most. And the children who do not need it, or don't need it yet, will benefit as well. For them, it could be helpful for later, or it may merely ensure that they maintain a positive way of dealing with their weight and health.'

She will now explain the details of the *universal prevention programme*, whose efficacy the Obesitas Research Institute is so eager to prove by means of sound scientific research. Then she sees that one of the teachers has raised his hand.

'Do you have a question?'

'Yes, I have. Are you simply trying to say that physical education is an important school subject?'

2. Camouflage and consciousness

The evolutionary struggle for survival has produced a fascinating phenomenon: the distinction between appearance and reality, between what something looks like and what it actually is. It can be found all over nature: camouflage. If you don't look like what you actually are, this can dramatically increase your chances of survival. Camouflage is not only a characteristic of prey animals such as stick insects, peacock butterflies and flatfish, but also of predators such as leopards and tigers, whose spots and stripes make it easier to ambush their prey. In a serious confrontation with a bird or a lizard, a stick insect won't stand a chance. And that is why it is so clever that the stick insect is not what it seems: it looks like a twig rather than a snack. If the deception is successful, the thrush will hop on, too hungry to see through the ruse. A similar but reversed situation can be found with the leopard. In a direct confrontation with an antelope, a leopard won't stand a chance either. It may have a high take-off speed, but its endurance and agility are second-rate, compared to an antelope. Add to this the antelope's natural vigilance and you will understand that the leopard needs the camouflage of its skin to approach his prey closely, so that during a surprise attack its take-off speed will be high enough.

In the evolutionary arms race, the development of camouflage undoubtedly went hand-in-hand with the development of a magnificent remedy against it: cognition.[1] Cognition is the ability to expose appearance as what it really is. If your cognitive abilities are developed well enough, you can see through the camouflage, and you can discover what is really in front of you. This is why birds also pay attention to movement, rather than to form alone. This is a shame for the stick insect, because it will be caught the moment it moves. Aha, the bird thinks, this is not a real twig but an insect: dinner time! Antelopes have developed a more sensitive nose, so that the leopard's skin has lost a great deal of its effectiveness. Despite those misleading spots, with its improved nose the antelope can smell the sweat of concentration of its attacker at a great distance. Of course, these are only small steps in a long arms race. Soon enough the leopard will develop a new, appropriate camouflage. It will stalk its prey against the wind, so that it won't be betrayed by its scent.

Where does this arms race end, this struggle between camouflage and disclosure? You may think it will not end at all. But since the Enlightenment there has been another answer to this question. The cognitive arms race ends in modern science. The – too suggestive – idea is the following (You will have to imagine a populist drum roll yourself, so that I can be safeguarded against the brazen chauvinist positivism that this idea exemplifies.). Nature has created human consciousness as the jewel in the crown of evolution. Our consciousness is really capable of discovering the definitive difference between appearance and reality. Our consciousness can really give us insight into reality. It cannot only discover what is actually the case, but also understand that appearance *is* appearance. Our consciousness can establish the essence of reality. This also means that it can determine the essence of appearance *as appearance*. Functionally, a bird can discover the difference between an ordinary twig and a stick insect. That is how it finds its food and how it stays alive. But the bird does not understand that the stick insect is trying to disguise itself. It has no idea what camouflage means, and the same is true for an antelope and a leopard: it is true for all animals. However smart they are and however good they are at camouflaging or at detecting, they are simply interested in eating and in avoiding being eaten. All they are interested in is how to survive, and not in knowledge itself. They are not interested in the essential nature of the prey, or in the essential nature of themselves, or in the essential nature of life. They are not interested in seeing through appearance as appearance. They are not interested in knowledge in itself. That is the prerogative of our own cognitive faculty, our consciousness.

It is this human consciousness that cannot only functionally process information, but can also evaluate this information. Only human beings have knowledge, only they know what they know. To conclude this train of thought: only human beings are capable of *science*.

It is characteristic of science that it is the part of our consciousness that allows us to see through appearance. However, science is also open and accessible and can be checked by everybody. A scientist is not a visionary, not a person who has a unique way of penetrating to the essence of being, but rather somebody who observes carefully, reasons logically and reports transparently. Science is one of humanity's public achievements; in theory, it is accessible to everybody who has consciousness. Science is an institution which can and should end the unfounded power of seers, prophets and their disciples. In this sense, science is really the appropriate throne of our consciousness. Science uses the real power of human consciousness to end the tyranny of two barbarian tendencies. On the one hand, science offers us the possibility to see through and thus conquer the brutal and irrational forces of blind nature. On the other hand, science offers us the possibility to see through the arbitrariness of the esoteric doctrines on which so-called visionaries and prophets base their power. Science is public and is able to disclose things. Science makes the world transparent. It clearly shows us the essence of the universe, which is then visible for everyone.

3. Science

Obviously, this positivist image of science is debatable. Of course it is a naive and excessively optimistic way of thinking. And of course it can be criticised from the points of view of methodology and of philosophy of science, and also of sociology and of politics. In fact, this is exactly what happened, especially in the second half of the previous century. However, this has not changed much in our everyday idea of science, of the world, and of our cognitive faculties. Our everyday idea can be easily captured in the following image of an atlas.

Scientists are a kind of explorer. Like the fearless sailors of yesteryear, they go on board craftily designed ships used to sail distant seas, looking for unknown shores. Originally they had to take great risks, for example because nobody really knew whether the earth was round or not. They might have been sailing straight into an endless abyss, might have reached the edge of the flat earth and then have plunged into everlasting oblivion.

However, times have changed. There is already so much that we do know. The earth is round, the faraway shores have been charted, all land has been identified and most is inhabited. Scientists now work at large, durable universities, guarded by critical supervisory bodies and financed by demanding authorities. Only metaphorically speaking do they sail distant seas; there is little personal risk involved. They still produce atlases, metaphorically: not of unknown shores but of mysterious brain processes, unknown chemical reactions, untraceable black holes, unpredictable psychological disorders and unimaginably small or complex mechanisms. They map more and more details, produce libraries of knowledge and have an increasing knowledge of the methods needed to approach their work. Scientists have developed such magnificent and sensitive observation techniques, have established such impressive laboratories and have built such powerful supercomputers that we are making progress at a dizzying pace in mapping both the smallest details and the largest complexities. Science has become a gap-fill exercise. The empty tomes in which we need to publish the very last maps are waiting on the shelf; before long we will know everything.

We can easily use this clear and simple image of a scientist – a kind of explorer producing maps of unknown continents – to outline a number of well-known forms of criticism of science. There is internal criticism: criticism of the ships used to sail the world as well as the shipping routes that are followed; criticism of compasses, sextants and protractors used to measure benchmarks; criticism of the scales, longitudes and latitudes used in the drawing; and so on. This is the criticism that scientists use for each other, the criticism that deals with the rationality that scientists apply when, although they agree on the ultimate goals – the seas to be sailed, the measuring instruments to be used and the atlas to be produced – they disagree on the methodology used to achieve these goals.[2]

In addition, there is also external criticism: criticism of the statement that a continent should actually be mapped by means of an atlas; criticism of the partisan choices for or against the seas to be sailed; criticism of the idea that we need an atlas at all; criticism of the authority that a prestigious atlas demands in our libraries even though it could not tell us anything more than a children's drawing. This is the criticism that historians of science, sociologists of science and post-modern philosophers of science use to attack science, criticism that from an external point of view regards the internal scientific rationality as a locally practised dogma. According to these critics, scientists are members of an esoteric sect, a powerful bastion, that under the guise of Truth with a capital T has developed a fetish for atlases and associated paraphernalia.[3]

I will not discuss this criticism here, and will stay away from the debate. I am not specifically pinpointing science as a societal arrangement that needs a severe reprimand. In this book I am primarily making a case for common sense, rather than against science. In terms of the atlas metaphor,

I would like to characterise the main theme of this book as follows: hurrah for atlases! There's nothing wrong with them; I like looking in atlases. I love dreaming away while looking at maps. But to understand that an atlas is an atlas, containing maps of faraway shores and unknown places, and that I can get to know these shores and places by studying an atlas, by using the atlas as an atlas, I need a great deal of common sense. And this common sense cannot be found in an atlas. This common sense precedes the atlas, surpasses it, and, when there is humaning to be done, grabs the atlas and puts it away again on the bookshelf. More specifically, this means two things, and I will limit myself here to a purely metaphorical illustration. First of all, there is no atlas of our common sense, nor can there be one. Our common sense is not a faraway shore that you can map scientifically, because our common sense is exactly the tool you need to deal with an atlas. Second, there cannot be an atlas of our humaning, the verb. Human behaviour cannot be mapped scientifically. Or, put more carefully: suppose we had an atlas of human behaviour. It would then be just as difficult to deal with that atlas as an atlas as it would be to learn how to practise doing your humaning to the best of your ability. Moreover, such an atlas would advise us to take another atlas, and another, and another until we are dizzy. That is why such an atlas would hinder us in our humaning, would distract us from what it is all about: you, me, us.

In this chapter I also touch upon a minor theme: our consciousness and its evolutionary explanation as a cognitive faculty, an ability to be open to a world that is trying to shake us awake, a world full of camouflage asking to be exposed. Our human consciousness makes science possible; that is fair enough. But that doesn't mean that science is the best expression of our human consciousness. It isn't.

4. Different route

She had not understood him, the PE teacher. And he hadn't understood her, that much was clear. He was not budging an inch. In hindsight, he was an unpleasant man. He didn't want to agree on anything, he kept bringing up his hobbyhorse and repeatedly held up the discussion by asking annoying questions.

She has to pay attention to the road now; it is busy. This always happens when the weather is so bad. The rush hour has started early today; it seems to be starting earlier every week. She decides not to go past the office but to go straight home.

The PE teacher had been asking unkind questions, pedantic questions. She had seen the others sigh and this had given her the support to remain patient, to keep her spirits up, to distance herself from his hostile behaviour. She was going to show him how a scientist deals with animosity. After all, she is a professional. She's not quite sure whether it had worked; she decides to give the principal a ring the next day, to see what he has to say about it.

Visibility is poor due to the pouring rain. It is really busy, and now it's getting dark, too.

Look, a building site. New buildings spring up so fast nowadays; she can't even remember seeing these offices being built, and now all the windows are lit, so they must have been occupied for a while. Strange, usually new offices first stay empty for some time. The company must have been really eager to move away from the city centre.

She needs to pay attention now. Old Mill Lane? What is this? Oh, there is a turning to the industrial estate here. She didn't know that. She hardly ever takes this road anymore. Last year, when she was still living in Garforth, she often took the A64. Must be quite useful to have a turning here, especially with all the lorries.

A car pulls out immediately in front of her. She reacts routinely, brakes and then checks her rear-view mirror because this is the best place to get in lane. In a moment she will reach the Bramham Roundabout where she will have to take a right.

But what is this? All of a sudden she recognises the sound barriers. These sound barriers are on the A1: she is on the A1. She feels really stupid; she has simply taken the route to the office, as if she was coming from Garforth. Of course, the turning, the new offices.

Imagine thinking they had been built so quickly, and on the A64, while all the time she was on the A1.

5. Rationality and internal coherence

It can happen to anybody: you're driving home, without paying attention, and you automatically drive to the office. It is quite understandable that your imagination runs riot about the speed at which new offices have been built, about the eagerness of the company to leave the city centre, about the usefulness of that new turning; and all the time you are unaware of where you are actually driving. Psychologists call this 'confabulation' or 'cognitive dissonance': incoming information that doesn't really suit your starting point is interpreted creatively and transformed into inform-ation that confirms your presuppositions. And this all happens completely on its own accord, at the front door of your consciousness, so to speak, and automatically, before you even notice. We have all been there. You can be *so* convinced that you know where you are. If you stop paying attention, you can easily develop a kind of tunnel vision. Are there offices here? My word, they must have built them really quickly. Is there a turning here? Well, that must be useful with all those lorries. And so on.

Of course, there is no malicious intent involved, no intentional self-deceit. Such a train of thought behind the wheel is rather understandable and sensible. If you are driving on the A64 and you know there aren't any offices along the road and yet you see some, then your memory must be playing tricks on you. Or these offices must have been built recently. What else could explain it? In fact, it is completely rational to combat cognitive dissonance, to restore the order in your head and in the world with the minimum amount of adjustment and correction. This is what we do all day long, often already at the level of our senses. For example, you will hear 'how about a cup of tea?', when all I have asked is 'owbout-cuptea?' Because why would I say something odd like 'owboutcuptea?' Such a correction is a simple improvement, and your corrections behind the wheel are very similar, cognitively speaking. It is just that I have built a misleading scenario around them this time, but usually the number of times that you don't know where you are is very small. And the number of times that you come across changes in your environment that can be explained by human interference is usually rather large. Often enough you do not even know that you are confabulating. And often enough your confabulations are completely correct.

If you start imagining possible explanations while driving on the A1, you could actually say that the rationality of your mental housekeeping is quite in order. Your actions are rather sensible, given the fact that you do not know that your point of departure is flawed. You can compare your rationality to the extraordinarily clever and sensible attempts by ancient Greek astronomers to correctly describe the movements of the celestial bodies, while they were unaware that they had in fact started from the wrong assumption, namely the mistaken idea that the earth was the unchanging centre of the universe. Their rationality, like yours, is a question of internal coherence. You are simply trying to keep as many obvious truths as possible upright. In the light of your conviction that you are travelling on the A64, you try to understand and find a place for these new facts. And you can do so easily. It happens quite naturally, almost without thinking. It seems sensible enough, at first sight. *Seems*, indeed.

In cases like these, your rationality is nothing more than pretence. The *internal* coherence of your mental housekeeping is accompanied by a lack of *external* correspondence.[4] What you think you are seeing does not correspond with what is actually the case. This could be explained by discussing different points of view. From your subjective, biased, internal point of view you are acting rationally, but from an objective, impartial, external point of view it is clear that you are acting in an irrational manner. Every correction and adjustment that you make only makes things worse: only makes you travel on the wrong motorway for longer and as a result it only takes longer for you to be home. It's just like the ancient Greek astronomers who made their description of the celestial bodies more and more complicated and as a result were more and more mistaken. It's just like someone doing a crossword puzzle who has filled in a long, yet incorrect word. It will go from bad to worse, because every new word that seems to fit will eventually turn out to be wrong. It would be best to rip it up and start again, and from an external perspective this is a simple piece of advice. But from the perspective of internally coherent mental housekeeping it becomes ever more difficult to accept this message.

This is not quite such a big deal in the car. It is clear that you are wrong, and that you are on the A1. There is no denying that, although we could imagine some pig-headed driver in such a scenario telling their passengers to shut up if they were to tell the driver that they are actually on the A1, that those offices have not been built recently at all, and that this turning has been there for years. Unfortunately for the driver, he will inevitably have to eat humble pie in such a scenario. He might start arguing, and try and involve God or an ideal observer, but that isn't necessary at all. Wrong is wrong and stupid is stupid.

It is quite different if you are a well-educated behavioural scientist, if you solemnly believe in intervention research and if you are terribly worried about the health of the country's overweight pupils. Then you really cannot relate to an unpleasant PE teacher who makes cynical remarks. Of course, his perspective is not objective or impartial. Of course he cannot lay claims to any external correspondence.

But are you sure that you can? Simply because you are a scientist?

6. Metamorphosis

When I was at secondary school, I was thrilled by the Romantic image of nature creating humanity as a form of existence in which Mother Nature herself could come into full bloom. I sort of imagined that this had all started with the Big Bang, the limitless void that filled itself within a fraction of a second with matter, with glowing celestial bodies, slowly cooling down circling clouds of gas, ice, rock. Cold, barren, empty planets endlessly following their orbits. One of these planets has a favourable temperature of between 0°C and 30°C, and on this planet there is water sloshing about. At a certain moment single-cell organisms cause something completely new to happen: life. I sort of imagined that after a while there would be endless green plains, grass blowing in the wind, animals scurrying around, fighting to the death, gene mutation after gene mutation, until after a long time a group of primates would come together in the savannah. Together, they need to take care of their vulnerable and dependent young, and to make sure everybody pulls their weight. At a certain moment they start murmuring, prattling, talking. This is something completely new, *language*, a phenomenon in which the difference can be expressed between what I say and what I do, between what I say to other people and what I say to myself. This is why we, human beings, have finally found a way to show each other the difference between appearance and reality, between camouflage and exposure, a distinction that had lain dormant all that time in the endless cycle of life and death, of eating and being eaten, of escape and capture. In the language in which we, human beings, have learned to feel at home, I could see something new come into existence, which would last forever. We sleep, we human beings. We really sleep, as I could at weekends when I was a schoolboy, until late in the afternoon, in a way that is categorically different from the zombie-like vegetating that is often seen in the rest of nature. Our sleep is fundamentally different, I thought, because it is in contrast to something new, to something completely different, something that has become forever possible in us, in human beings: waking up, awakening: *consciousness*.

As a schoolboy, I thought this was a wonderful image: the universe trying to wake up, trying to find itself, trying to be all-understanding, trying to become aware of itself. And the universe has selected us for that task. I would roam through the polders, coming to grips with the ridiculous careers advice I had been given by the school (poet or shepherd), reading Pascal's *Pensées*. I could imagine the whole thing: that my consciousness was one of these mirrors in which the universe was lovingly watching itself, silent and satisfied. This was the time when George Harrison was beginning to see the light in India, and in a neighbouring town I was accosted by a follower of Hare Krishna. In his long orange garment, in sandals, with a shaven head with a long tuft of hair that he was allowing to grow gracefully, he sold me an album called *Metamorphosis*. On the cover there was a delightful drawing of a human being in the form of a butterfly, one that had just emerged from its chrysalis and was ascending to the heavens, leaving behind it the empty chrysalis of the caterpillar it had once been.

A bit daft, I now think. But although I did not realise this at the time, in the early 1970s this image was surprisingly well-suited for what I then found a much more bourgeois and unacceptably daft image: the modern scientist who considers himself an heir to the Enlightenment. Since even science knew what to do with our unique human ability: consciousness. Science also appropriated human consciousness, demanding the starring role in the definitive exposure of appearance, in the definitive discovery of the truth, in the definitive awakening, in life that from then on would be lit by the bright light of the Theory of Everything. After all, it was science that had seriously pushed forward our rationality, that had drawn up criteria with which systematic research methods needed to comply, so that it could show us the way beyond the boundaries of our inner coherence to *external correspondence*. Also a bit daft, this limitless arrogance, don't you think?

7. Awake

The story of the sleepwalking murderer, Kenneth Parks, really captures the imagination (See e.g. Horn, 2004). How amazing that it really is possible: to get up from your settee, grab your car keys, drive to your in-laws who live 14 miles away, kill your mother-in-law with a kitchen knife and nearly kill your father-in-law too – and all this in a deep sleep, without realising, without waking up during this atrocious deed. This story seems inconceivable to us because we are so familiar with our sleep-wake rhythm of waking up and dozing off, with dreaming and dreamless sleep as well as with our actions and our experiences when we are awake. In this sleep-wake rhythm, the building blocks of Kenneth Parks' story obviously belong to the waking hours. Getting out of bed, grabbing a bunch of keys, opening the door, and driving a car: these are typically things that you do when you are awake. Stabbing a person is of course something you would never do, but if someone were to commit such an act, it would be done while the person was awake. Of course, sometimes you do things without paying attention, without realising – for example, you may at some point take the A1 to work even though you are convinced you are on the A64. But then you won't be asleep; it is just that you won't be as awake as you know you can be, as you know you sometimes *should* be when driving a car.

Despite the apparent evidence to the contrary, Kenneth Parks' story shows in a manner which is both intriguing and disquieting that sleeping and waking are not absolutely separate conditions; the distinction between the situation in which somebody is totally and completely awake and the situation in which somebody is totally and completely asleep is not as uncomplicated and crystal-clear as it might seem. Based on this idea, I would like to draw three conclusions about the meaning of our consciousness and the modern idea of the growth of our knowledge.

First of all, the dream of science – of the definitive awakening, of life in the bright light of the Theory of Everything, of the definitive liberation from the cycle of human suffering, of the Enlightenment – this is really only a *dream*, an idea that belongs in our sleep.[5] The idea of being totally and utterly awake is not at all clear and obvious and therefore not completely realistic. We do not have a well-determined idea at all of how it is possible and what it would mean to be totally and utterly awake. Of course, deep down we know this in our simple everyday lives. Throughout the day, waking and sleeping are intertwined in all kinds of fuzzy ways. There is the long cycle of 24 hours, in which you spend a substantial part in bed with an emphasis on sleep. However, there are also shorter cycles, of hours, minutes and seconds. Even when you are studying and trying to concentrate fully for 45 minutes, you drop off now and again. And this is also true for surgeons, referees and pilots. They need their breaks too: they need to let go for a while, to muse, to rest, to dream, to recharge. In a metaphorical sense, this is also true for people who are keen to recognise appearance as appearance, to unmask camouflage, and to confront the essence of being. Conscious creatures especially need the dark background that is necessary to highlight whatever or whoever is in the spotlight. In other words, if you want to recognise the camouflage of your prey *as camouflage*, the background needs to be as uncomplicated as possible. Knowing everything fully is an incoherent idea, a dream. And to have a dream, you need to go to sleep.

Second, if sleeping and waking are not absolutely separate conditions, this simple dichotomy is misleading. There are different ways of sleeping, just as there are different ways of being awake. In reality it is always a mixture: a little bit awake, or a little bit asleep. In the end, however, it is equally misleading to conclude that we are dealing with a continuum with two extremes, and this is not because we cannot conjure up an explicit image of these extremes, being utterly awake or utterly asleep, even though this inability does play a part. It is rather because for such a continuum we would need a quantification of components, so that we can determine whether a random situation in that continuum is closer to being completely awake or closer to being completely asleep: this would be equally misleading as the dichotomy.

To elucidate this conclusion, I will restrict myself here to the metaphorical meaning. Let's consider the optimistic idea that in the history of science there has been a permanent growth of our knowledge: we are becoming increasingly more awake. In prehistoric times people knew less than in the time of the ancient Greeks, which was when they knew less than in the Middle Ages, when they knew less than in the age of Darwin, when they knew less than in the days of the first computers, when they knew less than now. There is the obvious suggestion of linear development. However, it has been shown by many historians of science that this development happens in fits and starts and cannot be described in terms of how awake we are, or how much external correspondence we have reached. Philosophers of science nowadays do not think in terms of correspondence at all anymore, but rather in terms of increased coherence.

Coherence is a good idea, also in apparently clear examples of correspondence. Driving in your car on the A1, you must be incredibly stubborn if you won't admit that your passengers are right. In this case there is so much background information that you share with your passengers in an obvious and uncomplicated manner that it would really be irrationally foolish to continue upholding that you are actually driving on the A64. So you do not need any external correspondence at all to decide the matter, even though it seems that the external correspondence is decisive. *Seems*, that's right.

Let's suppose that you are solving an enormous crossword puzzle with a friend. The newspaper has offered a fantastic reward, so you have decided to fill it in together. You need to work with the clues in the newspaper, and with each other, of course. In the middle of the crossword there is room for a nine-letter word. The clue is given as 'the power of our consciousness'.

— Depiction, your friend says. Nine letters, it has to be right.

You're not so sure. And then suddenly you see the light.

— Coherence. *That's* what it must be! Nine letters. Of course.

This is also the way things work in science, even though there is no crossword editor, no one with a genuinely external perspective.[6] No one who is going to tell you whether you have won or not. You live your life in a good or a bad manner, for a long or a short time, in the light or in the dark. That is all. Together you and your friend fill in the rest of the crossword puzzle. You start with the word 'coherence', because your friend is not so argumentative. But after a while you get stuck. Perhaps it should have been 'depiction' after all. Start again? Go on for a bit? In the words of the metaphor: how awake are you? How many words do you think are correct? What makes you so sure? Or to put it differently: how awake was Newton when he thought that space and time were absolute dimensions? How certain was he that there is absolute time? It seems like the wrong question. *Seems?*

There is a third conclusion. Although being awake is a relative idea, this doesn't mean that there is no difference between waking and sleeping. The crux is the word *being*: it is the wrong verb. It is not about *being awake*, but about *waking up*. In our consciousness it is not about *being* aware, but about *becoming* aware. It is about the action rather than the situation. It is the same with the verb humaning: it is a matter of doing rather than of being. It is about that short moment when you see how things really are, the *Aha* experience, the moment in which the universe is positively disposed towards you and offers you a window to look inside, shows you the camouflage *as camouflage*, so that you can see how things really are. There, in that scenario, at that particular moment.

We have cognitive skills. We have consciousness. We can unmask camouflage, and we can learn to distinguish between appearances and reality. We can do so individually, situation by situation, moment by moment. You may well think that you can add these things up, that you know more and more, that you have unmasked more and more false appearances and as a result have gained more and more knowledge. You may think that there is a linear growth of knowledge, an increased understanding, and libraries full of atlases. You may well think that you are peeling an onion, layer by layer, until you come to the essence and find yourself eye to eye with the real hard core, the essence of being, the Truth with a capital T. It is a mistake to think in such a static manner, as if you can collect consciousness, like beetles in a box. As if there won't be an additional layer under the last layer that you have peeled. As if it is not about the peeling itself but about what you find under each layer. As if you can simply keep all that knowledge in atlases, in stuffy libraries and never need to look at it again. Because now you know. It's all been archived. Safely stowed. As if it is not about the effort of studying a map again, not about the *Aha* experience that lies in wait for you in the atlas if you are lucky, not about that PE teacher who has made you suddenly realise that universal prevention is just another word for education, for upbringing, for development: words whose sounds we have recognised ever since we started to murmur them over 100,000 years ago. In those days we were not trying to prevent obesity but child mortality and hunger. Was anybody talking about progress? Is anybody waking up?

8. Now what?

If you use your common sense, you do many things on autopilot. Of course, it is possible that people around you have been trying to fool your passengers, that they have ingeniously placed a film set along the A1, make-believe fronts pretending that the offices on the A64 are now there; it is certainly possible that they have used painters and photographers, have spared no expense to produce a gigantic poster which gives the impression that this is the turning for Old Mill Lane. It is all possible, but if you need to take such scenarios into account, you will never get anything done. Who knows: they may have collected a group of major actors who act as members of your household. Who knows: you may have been hooked up to a computer since birth, like a brain in a vat, and all your experiences are simply the effect of a virtual reality program linked to your brain. It may also be less drastic: perhaps the PE teacher has been asked by your employer to come up with a difficult question in order to see whether you can function as a loyal employee of the Obesitas Research Institute, even under difficult circumstances.

Of course, it is all possible: but it is so unlikely, ridiculously unlikely. That is why you just do not take such exceptional circumstances into account. After all, you use your common sense. You simply trust your folk physics and your folk psychology. In other words, you simply assume that your everyday expectations are appropriate, that the world is sticking to its usual routine and that the internal rationality of your mental housekeeping is in order; therefore, you can simply assume that the obvious, unobtrusive background does not need any special inspection. You do not permanently check everything like a scientist, there is no need to check all the variables, or to observe everything carefully and in a methodologically sound manner; instead, you simply assume that the things that in your opinion are understandable, efficient and good are in fact just that: understandable, efficient and good. You simply go with the everyday flow, just like everybody else. There is no expertise and no science involved. You're conscious, of course, but this is the usual mix of longer and shorter wake-sleep cycles, of drifting off and waking up.

However, you get something extra if you use your common sense, if you try to do your humaning in the everyday flow to the best of your ability. First of all, you trust your fellow humans to warn you when you sink too deeply into a dogmatic slumber. Second, you are accommodating: you give other people the scope to make a case for their own ideas and perspectives. After all, you are conscious together, with each other. It is perfectly possible that the PE teacher notices something that you have overlooked because you think of overweight people all day, of health risks and of interventions, of deliberate, scientifically sound interventions in scenarios that are perfectly normal to other people. And third, crucially, you are aware of grey areas. You want to understand yourself and your reactions to your environment. You pay attention to the questions that might pop up when you find yourself confronted by potential ambiguity. You are wary of camouflage, of your own emotions and those of other people, and of the possibility that the obvious only seems to be obvious. Seems.

It is a matter of sensitivity, of your humaning consciousness, of a certain awareness of the irony that can present itself in the guise of a sound barrier that you didn't expect to be there, or in the guise of a PE teacher asking the right question.

Notes

1   For a detailed inquiry into this evolutionary origin of cognition see Sterelny (2003).
2   A remarkably crisp discussion of the methodological concerns of social science can be found in Macdonald and Pettit (1981).
3   Much has been written about these so-called "science wars". See Parsons (2003) for a clear overview.
4   The theme is a well-known one, and very clearly discussed in Davidson (1985).
5   The metaphor is Stanley Rosen's. See Rosen (1979), pp. 149–156.
6   The metaphor is Susan Haack's. See Haack (2009).

8   Understanding without objectivity

1. Hockey coach

His suit fits perfectly. He looks great. Soft, elegant fabric with a beautiful deep colour. Dark grey. He always wears a shirt as white as snow, top button undone. No tie. And how about the shoes? Light brown, sharply cut, fine leather. Italian design, of course. It makes them look both smooth and sophisticated. He looks . . . suave.

When he looks at himself in the mirror, about 2 hours before the match, he understands why the girls like him so much. It wouldn't surprise him if one or two were secretly in love with him. His hair is perfect too. Blonde locks, carelessly combed back, perfectly matching his bronzed skin.

Of course, it is his training methods and his coaching ability that count. They will make this team successful. But it is obvious that his charm and charisma are equally important. These will drive the girls to the limit of their abilities. When he is standing on the field – always standing, you won't catch him sitting in the dugout – when he is standing on the field, one hand loosely in his trouser pocket, aware of the spectators watching him coach – when he is standing on the field, overseeing the match, making a well-placed remark here and there, he knows that these girls will go through hell and back for him. They will give their all to win. For him.

They have brought him to *Penzance Pirates HC* to make this team a success. For him it came just at the right time; he was ready for a new challenge, as he didn't like the way things were going at *West Grinstead HC*. The people there were getting on his nerves, so it was good that there was an opening here. It was good that they could clearly see the value of his input for this team, for the future. Because that is what it is all about: that you can see something in a team, that he can see something in this team, and that *Penzance Pirates* can see something in him.

2. Response-dependent properties

One of the most fascinating permanent questions in philosophy is what to think of our own experiences. It all starts quite simply: you are beside a hockey pitch watching a team play in a way that makes you think, 'Wow, that team plays together well!' And if hockey – or sport in general – isn't your thing, choose your own favourite example. Listen to music and hear a sparkling riff, have a look at your finest glass statuette, smell the scent of a strong espresso, or make it almost banal and feel the coarseness of a towel washed without fabric softener. It really doesn't matter; every experience can be used to introduce this issue. Feel the weight of a steel padlock, smell the perfume of a mature woman, or watch a vain hockey coach admire himself in the mirror.

Whatever experience you start with, it is always the case that a certain content or quality presents itself to you. This content is accompanied by on the one hand the realisation that there is something 'out there', and on the other hand the realisation that you are the person who realises this now. Traditionally, philosophers use the concepts OBJECT and SUBJECT to refer to these two aspects of experience. In experience subject and object meet, which inevitably leads to the question of how much the object contributes to the experience and how much the subject contributes. Feel that coarse towel and realise that the towel which is 'out there', hanging on the washing line, meets you in this feeling, and then wonder whether that coarseness is a quality of the towel, or rather a quality of your skin, your sense of touch, or maybe even your brain. Or does this coarseness only exist in the meeting and is it meaningless to think about the coarseness as being *there*, in the towel, or *here* in your skin? Try and experience it as a phenomenologist. Ignore all the knowledge that you have of fabric softener, towels, washing lines and so on. If it helps: imagine that you are a new-born baby. Be completely absorbed in that one experience, exclude all knowledge other than the fact that you are now completely immersed in the one sensation that is everywhere: *coarse*. How coarse this feels!

Then wonder how much information this experience gives you about the world and how much about yourself. If this example doesn't tickle your fancy because you have no aversion to coarse towels, imagine yourself blue in the face with laughter or moved to tears, from watching your favourite stand-up comedian or your favourite tragicomedy. Now ask yourself the same question: is that stand-up comedian really so funny, or is your laughter caused by your (substandard?) sense of humour? Is that film really so moving or are you just a sentimental soul? Of course this is not the only question we can ask if we want to know what to think of the content of our experience. But this seems to be a good first question because it encourages us to think about the meaning of the fact that we meet reality in our experience. What exactly is it that comes together when a hockey coach sees his team play well together?

Nowadays philosophers like to talk and write about 'response-dependent properties' to get a grip on what is going on when we meet reality in our experience (See Casati & Tappolet, 1998). What they say is that some properties are only qualities of the things in the world in virtue of our response to being affected by these things. The best-known example in the history of philosophy is doubtless the red colour of a ripe tomato. By now science has made it clear to us that the tomato is not really red, at least not in itself. If you were to place a tomato in complete quarantine, somewhere in an isolated corner of the universe, and you didn't allow any light or any creatures such as us with eyes sensitive to light with a wavelength of 690 nm, this tomato wouldn't be red at all. The tomato obtains its red colour from the cooperation between the light and the cones in our retina. In a way this beautiful red colour is a present that our eyes give to a lovely, ripe tomato. If we are not involved, the tomato is not red at all; all it does is reflect light with a wavelength of 690 nm.

This idea has become common knowledge, I think, and will probably continue to hold. It is the robust result of the impressive empirical tradition built on Locke's distinction between primary and secondary qualities. Even so, some tricky problems have been associated with this idea, issues raised by Wittgenstein and by other philosophers who have continued his line of thought. These problems are not so much related to the specific, personal, subjective phenomenology of experiencing 'red'. Of course, I do not know how you experience the red colour of a ripe tomato. For all I know, my eyes may be completely atypical. Other people might be completely distraught if they were to see the world through my eyes for a day. Such colours! Such variety! And such terribly clashing contrasts! Who knows, I may indeed see all the colours completely differently to other people. Or perhaps you do. That is possible, but as long as our reactions to different colours systematically agree with those of other people, it is no great matter. If we – just like everybody else – immediately recognise a tomato in a heap of orange satsumas and recognise that the colour of the tomato is the same as an old-fashioned phone box and a fire engine, and we can enjoy or abhor the painting *Who's Afraid of Red, Yellow and Blue* by Barnett Newman in a way similar to other people, it becomes useless, according to Wittgenstein, to worry about the possibility that my experience, phenomenologically, may be completely different to yours.

However, there are further questions. If it is structural that our reactions do not agree, to what extent is this acceptable? Of course, blind people are a case in point, just like deaf people. Their reactions do not agree at all with those of other people, but theoretically that situation is easy to deal with. If a person's senses have been damaged, it is obvious that we cannot expect a reliable structure in their responses. In a more complex but still understandable manner, the same also holds for colour-blind people. The fact that we experience light with a wavelength of 690 nm as red is the reliable result of a correctly functioning visual detection organ. But isn't this a slippery slope? When do people have correctly functioning senses? And how about a colour like turquoise, which many people think is a kind of blue, but an equal number think is some sort of green. And how about the variability in other modalities? When is somebody talking too loudly? When too softly? To what extent can you say that hairspray smells awful? How coarse is coarse? Once you start thinking about these issues, you discover something fascinating: there is a clear structure in our responses! But is that structure dependent on contingent factors that may well be manipulated by us? For example, is that structure a question of habit, upbringing or origin?

It is this issue that Wittgenstein puts centre stage. According to Wittgenstein, the response that gives secondary qualities their specific character is not so much a subjective, phenomenological response but a *conceptual* response (See Wittgenstein, 1977). It is only because I have acquired the concept RED that I respond in a correct manner to the colour of ripe tomatoes, fire engines, old-fashioned phone boxes and paintings by Barnett Newman. And I acquired that concept when I was young, and slowly but surely became a member of our language community. I could have completely failed to acquire the concept RED, or could only have acquired it partly, if I had been blind. But incidentally, even if you are blind, you still have some idea of the concept RED. After all, other people use this concept and you will pick up some elements of the structure that you notice in their language use. Even if you are blind you will know that red is a colour, that it is to do with light, that it is a way to experience the surface of objects. In that sense to a blind person RED may be similar to the concepts COARSE and SMOOTH.

Response-dependent concepts form a much larger class than just the concepts directly related to our five senses. This is what makes the issue so interesting and so important. You can start by thinking of aesthetic qualities, such as beautiful, ugly, lively, exciting, intense, harmonious or disgusting. And then think of ethical qualities, such as good, dishonest, just, wicked or evil. These are all qualities that are related to response-dependent concepts. And now you will feel the issue shift, because what should you think of qualities that you really need to learn to recognise and distinguish by lingering in specialist domains for a sufficient amount of time. For example, when is an action on the football pitch a foul, and when is it a correct tackle? When does the sound of an engine ticking over indicate that there is something wrong with one of the cylinders? When does a dark spot on an x-ray indicate metastasis? How many of such spots do you need to see before you can conclude that you have acquired the concept METASTASIS? And even then . . . how many endless discussions can you manage before you or somebody else starts wondering whether you have actually mastered the concept METASTASIS? If we look at football commentators on TV, it seems that they never agree; still, they all appear to accept that individually they know what they are talking about, namely about the game and the football players, about the reality in the stadium.

Of course it is possible that there is no structure in matters related to football, but there is clearly a limit to the possible chaos. And that limit cannot be found on the football pitch, or in the columns of *World Soccer*, or in the good-natured banter on *Match of the Day*. Those limits are located in a totally different but extraordinarily interesting place: in the ability that children (or, more generally, newcomers) have to acquire language. If you had to learn football concepts from a group of commentators who constantly disagreed and never used any concept in a coherent manner, you would remain confused and would never be able to join in the conversation. Imagine what would happen to a toddler trying to acquire the concept of RED, when surrounded by adults who point at red, blue, green, brown or yellow objects in a completely unpredictable manner whenever they or the toddler utter the word 'red'. This particular toddler will never learn to use the words for the colours correctly, no matter how much coherence and structure he is able to distinguish in the phenomenology of his colour experiences.

Can you see what this reasoning leads to? Response-dependent properties – the properties that objects are given by us, by the coherent manner in which we react to these objects – are related to response-dependent concepts. These are the concepts that we as members of a language community learn to use in a coherent manner when we learn to talk about how we engage with each other and the world. As a result, it is in the language, and in the harmony of the language community, that we and the world meet. This harmony is a complex and dynamic topic, which is sometimes characterised by an exceptional fragility and sometimes by an extraordinary sturdiness. It sometimes seems like an undeniable fact, a robust evolutionary result, but sometimes also surprisingly flexible and pliable, a youthful thing cooing and crowing new concepts into existence, such as HELICOPTER PARENT, BROMANCE, CROWDSOURCING and SEXTING. Meeting the world is not a question of living through isolated, subjective experiences, but a matter of being able to live in the language, of cooperating towards a shared understanding, of living together in a harmonious language community.

And there is another effect. In the eighteenth century, Berkeley and Hume showed that Locke's distinction between primary and secondary qualities is not easy to maintain. They argued that what holds for secondary qualities may well hold for all qualities. In other words, objects in the world may have all their properties only because of the contributions made by subjects. This conclusion was lifted to a higher level and drastically transformed, first by Kant and then by Hegel. Following Wittgenstein, we can now formulate it in a manner that is again different. All the concepts that we use to represent the content of our experience may well be response-dependent, which means that there may not be a strong, objective and concrete distinction between on the one hand seemingly subjective concepts such as COARSE, SPARKLING, BEAUTIFUL, RED and CHARISMATIC, and on the other hand seemingly objective concepts such as MASS, TEMPERATURE, NEUTRINO, TELOMERE, CONFIRMATION BIAS and HYPOTHALAMUS. In all these cases, the structure in our conceptual responses may eventually be the accomplishment of our ability to form one language community, rather than the inevitable effect of the physical processes that we are subject to in this universe.[1] This sounds rather blunt. Aren't we going to lose objective reality in this manner?

3. Affordances and our language community

Of course, it is not by some obscure miracle that we have formed a language community with the people around us. What else could we have done? Still, there is a long and honourable tradition with an important message that wants us to believe that it is a wonderful phenomenon that we can understand each other so easily and so well. This tradition culminates in Descartes' first meditation, in which he unflinchingly tries to doubt anything that can possibly be doubted. He becomes completely isolated. Thrown back upon his own resources and locked within his own mind, Descartes realises that there is only one thing that he cannot doubt: the fact that he's a doubting soul.

Try and look at this in two different ways. First of all, suppose you are the person undertaking this meditation. Sit yourself down in a corner of your room, close the curtains and imagine that there is nothing at all outside the room. The universe is empty; no matter which direction or how far you look, there is absolutely nothing to see, to hear, to smell or to feel. Nothing. Emptiness everywhere. Your room is hovering somewhere in the enormous void of the universe as a bizarre backdrop to this meditation. Right. Now close your eyes and also imagine the room gone. Concentrate on nothing at all, or, if that is too difficult (I never seem to manage this), concentrate on your breathing. That is all there is; there is nothing else. No curtains, no walls, no furniture, no floor, nothing. The hardest part is yet to come, but if you have made it this far, you will also manage the last part. Forget your body. Forget that you are a body. Forget that you have a body. You haven't felt the floor for a while, but now you no longer feel your back either, nor the position of your feet, your legs, your arms, your stomach, your head. Nothing. Go completely inside yourself. Pretend that this means something. Turn your back on everything. And be. Just be. Be completely alone with yourself. With yourself and in yourself, a completely isolated and solitary soul. If Descartes is right, you will now notice that in this simple, naked being, as a soul, you cannot be anything else than a thinking something. Your being is a question of thinking. Hence Descartes' famous adage: *I think, therefore I am.*

Right. Now it's time for the other way. Listen to yourself concluding that there is one thing that you cannot doubt, namely that you are a doubting soul. Huh? What is that: doubt? And soul? How can you picture that? Where do these words come from? Words? What do they mean? What are words? What have they got to say? Why is it that I have these words, that I know what they mean, and that I know that I am indeed 'a doubting soul'? Do you see, this way is much faster. Look at yourself, sitting in the corner of the room, eyes closed. Look at yourself following my directions, trying to get inside yourself, trying to end up completely alone with yourself. Now determine how great your command of the language must be to follow my directions, to undertake Descartes' meditation for yourself, to be able to imitate him. Realise how much language you share with me, need to share with me, just to be able to meditate. Realise how intensely you must be a member of our language community before it is even possible to arrive at the place where Descartes thinks that it all starts. If you were to renounce this language community, if you were to exclude our shared language by means of Descartes' radical doubts, you would not even be able to arrive at the place where he thinks that it all starts.[2] And a doubting soul: how would you ever be able to know what that was; indeed, how would you ever be able to be a doubting soul, if you didn't take part in our language?

This rhetorical question points towards what Wittgenstein is trying to tell us. Our individual experiences do not precede our common language, neither logically nor chronologically. It is not true that there is first an experience that you then need to find words for; it is the other way around. First there is our common language: fragile, sturdy, ambiguous and self-evident. This language gives us our experiences: the coarseness of a towel, the beautiful, deep red colour of a ripe tomato, the sparkling rhythm of a riff, a stand-up comedian's brilliant humour and the wonderful teamwork of a hockey team.

Let's look at the way it all begins. You're in your cradle, newly born. You are surrounded by scents, colours, warmth, soft caring hands, calming sounds. This is a pure, inarticulate existence, merely a state of mind. You hear whispering, friendly voices. The language that your father and mother share already roams free outside of you. There is kindness, tenderness, encouragement, enthusiasm. Everything is full of love. Mumbled phrases, loving sentences, affectionate remarks. You don't understand a word, but you wallow in the affection, enjoy the care and the attention, flourish in the rhythm of ever-recurring vocal sounds. And slowly but surely, unhurriedly but clearly, patterns are formed and connections are made. You start to recognise your mother's voice and your father's voice. Ever-repeated sounds are becoming clearer. You start to react to your name. You get used to the love that accompanies the two syllable sounds that are repeated over and over: Mummy. Daddy. Mummy. You pick it up. Actually it is incredible how fast you pick it up, how fast you get used to the rhythm and diction, to the vowels and syllables that are characteristic of your mother tongue. It is almost a miracle how easily everything comes together, your parents' voices and the clever alertness of your language capacity. If you look at it from a distance it is equally miraculous, equally characteristic, and also equally natural, as the ease with which a young starling opens its beak, a duckling starts paddling in the water, and a foal starts frolicking in the field during the early days of spring.

There is a word for this, originally introduced by J.J. Gibson (1979), which nowadays often emerges among scientists and philosophers who are thinking about cognition and who have become enthusiastic about what seems to be a new paradigm. This paradigm is based on the assumption that knowledge is embodied, and develops locally, in the spontaneous activity of creatures who have an interest in their surroundings. The word is *affordance*, and refers to the ability of your environment to offer you something, to invite you, to give you the opportunity to react in a specific manner. This manner matches the characteristic features of the environment, but at the same time, *emphatically*, also the characteristic features of the kind of creature that you are. If a wolf meets a goat, it sees a tasty morsel. It is the goat who presents the following to the wolf: 'Look at me, I am here to be eaten!' In contrast, a goat that meets a wolf sees danger. This is what the wolf shows the goat: 'Careful, I'm a creature to be afraid of!' However, a wolf meeting a cabbage simply looks the other way. He is not interested in the cabbage. But in fact, this is a missed opportunity because the cabbage embodies a surprising *affordance* for the wolf. After all, the cabbage can be used as 'bait for goats', as goats see this as a tasty morsel. It is fortunate for the goat that the wolf doesn't understand this. For him the cabbage is simply an obstacle, something in the way, an *affordance* 'to jump over'. For Wayne Rooney and Lionel Messi, however, this same cabbage would be something different, an *affordance* 'to kick'. For them, most things are made to kick. In short, it all depends on what you see in things and at the same time, it also depends on what these things offer you, what they invite you to do or give you the opportunity to do.

AFFORDANCE is a thought-provoking concept. You might think that this concept shows us the way in the complex and obscure jungle of response-dependent properties. It works as follows: not everything that rolls is meant to play hockey with. You can't play hockey with a car tyre or with a tomato. Perhaps you can with a cabbage, for a while. But you definitely can with a hockey ball, for a much longer time, because it is actually meant for that; it really has the affordance 'to play hockey with'. In a similar way you could say that the world has affordances to activate certain ideas. You could react to a tomato by describing it as red. A tomato has the affordance 'to describe as red', just like an old-fashioned telephone box and a fire engine. It seems that all the surfaces that reflect light with a wavelength of 690 nm have the affordance 'to describe as red'. The knowledge that we have acquired as human beings has provided us with this insight, and it is this knowledge that maps a great number of affordances: surfaces that reflect light with a wavelength of 470 nm are blue; yellow belongs to a wavelength of 580 nm and violet to a wavelength of 400 nm. Colour theory has mapped this knowledge and has anchored the meaning of the different colour words in the affordances of surfaces. This could be expanded further. After all, we know that after washing a towel, salt crystals are left behind and these form a thin layer when a towel is dried on the washing line. It is this thin layer that makes the towel feel coarse. The affordance 'coarse' comes with the salt crystals. And you could go on like this: not everybody is charismatic, but you could look into the shared characteristics of Bill Clinton, the Dalai Lama, Oprah Winfrey and Jim Morrison, in order to find out if your hockey coach has charisma. If he is charismatic, it means that he has the affordance that invites you to react in a similar way as you would do to Bill Clinton and the Dalai Lama: the affordance 'to revere', 'to be enthused by', 'to go through hell and back for', and 'to listen to in captivation' and then also 'to describe as charismatic'.

A concept is not simply a word; a concept is the correct way of dealing with a word. If you master the concept CHARISMA, or the concepts RED, COARSE, HYPOTHALAMUS or BAIT FOR GOATS, you can use these words in the correct manner. You know when to say what. The suggestion of the reasoning of this section is that the correct usage of a word is a function of the affordance that a certain environment has for you, to give you the opportunity to come up with that particular word. For example, when you see a can of coke lying on the pavement, you may feel inclined to give it a good old kick because such a can has the affordance 'to kick'. But you can also feel something else emerge at the same time, namely the word 'can', because such a can also has the affordance 'to say "can" to'. If you let this sink in, you will see that there are two routes that you can take to go from an observed can to the concept of CAN. The first route is from the thing to the word; the second goes from the word to the thing. I will use two examples to try to explain this.

Let's go back to yourself as a baby in your cradle. Realise how intangible your experience is: it is a comprehensive, unstructured atmosphere, a confusion of impressions, an ebb and flow of feelings: warmth, love, desertion, hunger, security. The patterns that slowly show themselves, the coherence in the indulgence given by your mother and father go together with the coherence of the sounds that you hear: Mummy, Daddy. These are words that you will learn to use in time, words that show you the way to what a mother and a father is, just like your own mother and father have shown you the way to these words. Now also try to imagine how you acquired the concept RED: encouraged by your father and mother, or your siblings, you slowly understood that the red building blocks, balls, keys, cars and pieces of cloth were related and belonged to that word 'red'. The spoken word, 'look: red', showed you the way to these things even before the things showed you the way to that word. In other words, it is both the environment of a language community and the environment of the things that provide you with affordances.

And now for the second example. You may never have regarded a cabbage as bait for goats. In fact, it is a rather peculiar way of looking at it. But in the context of the well-known story of the wolf, the cabbage and the goat that I implicitly referred to earlier in this section, it is a phrase whose meaning you immediately grasp. Straightaway, the phrase provides you with a new concept: BAIT FOR GOATS. The phrase invites you to look at a cabbage in a new light. The phrase has an affordance; it gives you the opportunity to see something new in the cabbage. That new element itself is also an affordance, a characteristic of cabbage that cabbage has always had, even though it was never made obvious. Put a cabbage in an environment and that environment invites goats to appear on stage. This is very useful for a wolf who could do with a tasty morsel, but the message here is aimed at you, reader. I would like you to realise that people do not only live among things, but that they live their lives in a linguistic environment, in a language community, among words and concepts. Imagine the wolf putting a cabbage down somewhere, and watch how the cabbage entices a goat. See how it all comes together, how one thing follows from the other. Realise that you can discuss this interaction and can think about it with the concept of AFFORDANCE. Realise that the cabbage has an interesting, unexpected affordance for the wolf, because the cabbage is bait for goats. But you should especially realise that it is thanks to that word, thanks to the text that you are reading now, that you have discovered this affordance. And realise what this means: it is your linguistic environment, this book, that has affordances, that provides you with the opportunity to think about the interaction between a wolf, a cabbage and a goat. That is Wittgenstein's message. It doesn't start with an experience for which you subsequently find words. It is the other way around: our common language comes first and forms our experience in its wake.

4. Normal language skills and favourable circumstances

Let's sum up my story so far. In our experiences we meet the world. In these experiences things appear to have response-dependent properties. A ripe tomato has its beautiful colour as a result of the way our eyes react to the red affordance that the skin of the tomato has in full light. Response-dependent properties are related to response-dependent concepts. Because I know the concept RED, I can react to the tomato's red affordance in the correct manner. I have learnt this concept because I live in a linguistic environment, because I am a member of a language community; this has taught me to react correctly to the language affordances that are offered to me in the words of other people.

Other people's words are affordances for me, and this discloses two divisive complications that lie silently and unnoticed under the surface of our language community. They are not clearly marked and cordoned off, like unexploded World War II bombs in a farmer's field. In our daily lives we are only faintly aware of these complications, even though they receive ample attention from philosophers who occupy themselves with response-dependent concepts. One of the complications that I am alluding to refers to the language skills of the people around us, the people who gave us our linguistic affordances. The other complication relates to the circumstances in which we are confronted with these linguistic performances. I will elucidate these concepts one by one.

As a baby or toddler learning to speak, I expect the people around me to have normal language skills. This is something I *have to* expect. After all, these are the people who give me my language. These are the people who initiate me in the language community, who familiarise me with the words of my native tongue and their meanings. They have to be normal, in my eyes, and in this case both 'normal' and 'what I'm used to' mean 'what is adequately attuned to the language in question'. After all, it is their language community in which I need to feel at home. Of course their language skills are normal; the role they play in my life makes them exemplary representatives of their language community. These are the people who teach me what is red, and what is blue, coarse and beautiful. They teach me what a ball is, a can, a wolf, a cabbage, a sparkling riff and a hockey coach. If they do not know what they are talking about, how will I ever be able to learn? So the bare fact that I learn to speak, that I will eventually understand them, that I learn to say 'mummy' and 'daddy', and that even before I can talk I can proudly fetch my cuddly toy from the cupboard if they ask for it . . . these facts inevitably imply that their language skills are normal. It doesn't matter whether they are teaching me Abkhaz, Mandarin Chinese, Celtic, Portuguese or English; they know what they are talking about and I adopt their ways: their accent, their dialect, their words and their concepts.

The first complication that lies like a ticking time bomb under my language skills is the fact that I need to accept that 'what is adequately attuned to our language' is identical to 'what I am used to'. We always learn our mother tongue from a very limited number of people: our parents and siblings, extended family, a few friends and perhaps later some other acquaintances. None of these people may play hockey, or perhaps they all play it; no one may teach me the words for colour, or they may all keep on pestering me with building blocks in primary colours all the time; they may be mad about glass statuettes, play ear-piercing Metallica riffs all day, or hold interminable discussions about the new education reform bill. They may pray five times a day facing East and continuously drink mint tea. All these things are possible, and they will all be perfectly normal to me. No matter who teaches me my first language, it will take a very long time before I find myself in a situation in which I can begin to realise that the behaviour of the people I grew up with – which is the most normal behaviour I can imagine – may in fact be rather odd. It will be difficult for me to believe that other people do things differently. But once I have found myself in such a situation and have noticed with my own ears that not everybody listens to Metallica all day, but people may also listen to Bach or Robbie Williams, to radio talk shows or to nothing at all . . . once I start to realise that my family isn't quite as normal as I thought, the language of my environment has already lodged itself in my experience, in my way of thinking and in my way of speaking. I will never be able to get rid of this first language, however abnormal it may turn out to be.

And even though this first language will forever be my language, this doesn't mean that I am hopelessly locked in the thinking frame of this obscure and abnormal sect that has taught me how to speak (my parental family). For a language essentially has an open character. Once you have learned how to speak, even if it is in Mandarin Chinese with a heavy Celtic or Abkhaz accent, this gives you access to a world of concepts that is not limited to natural languages. For example, you and I might meet, while you are mumbling in Mandarin Chinese and I don't understand a word. Then this lack of understanding is what I'll be trying to make clear to you, using gestures. Depending on our time and energy, and also on the necessity to understand each other, we will persevere in trying to understand one another. Imagine us being locked up together in a henhouse, or in a room for 12 years, or meeting by accident on an uninhabited island. Despite your Mandarin Chinese, I will use my eyes and ears to pick up the affordances that you use to invite me into your world, just as you will use your eyes and ears to notice the affordances that I offer you in my best English.

Fortunately the differences are usually not as dramatic as in this example. Still, even if we both speak English, but I grew up among farmers in a small Yorkshire village and you in a back street in Harlem, we will also both discover that there is a difference between what we are used to and what is needed for an adequate attunement to understand our English language correctly. On the one hand this difference will cause misunderstanding, but on the other hand it will offer new possibilities of understanding. The misunderstanding is a result of the fact that in our use of language we always need to start with what we are used to; as a consequence, we are not adequately attuned to the other person because there is usually little that we share. Shining through this misunderstanding, however, new possibilities of understanding appear, because it is in this contrast that we discover some scope for a different interpretation than what is normal to us: an adequate attunement to our common language.

The other complication refers to the circumstances in which I learn to react to linguistic affordances. Suppose that I have learnt to use the words for colours in a room in which only a blue light has been shining. For all I know, tomatoes are grey and aubergines blue. The circumstances would not be favourable for gaining a correct understanding of RED, PURPLE, BLUE and so on. Or suppose that I grew up in an environment in which all dogs are either Great Danes or Yorkshire terriers; this would not be very favourable for learning the correct concept of DOG. Of course, we can think of endless variations of this second complication. This complication conveys that response-dependent concepts can only be acquired in a correct manner if the circumstances are favourable; in other words, the available affordances in these circumstances should be exemplary invitations and really give the learner the opportunity to develop the correct response.

Let's suppose that people used to swear a great deal in my family. Suppose that in my family it was easy to notice the inflation in meaning that you can also observe in society nowadays – that you need to explode rather dramatically so as to communicate even a small amount of indignation. Let's suppose that it was normal in my family to use three or four of the worst swear words if somebody accidentally bumped into you. And let's suppose I had met this girl whose parents were fanatical members of the No Cussing Club. It would be a terrible shock to her system if she accidentally knocked over a glass of lemonade at my home. I hope my point is clear. 'Sorry, my mistake' doesn't have the same meaning in every environment, just like the concept RICH means something completely different in the UK than in Somalia or in Silicon Valley. Nowadays, a large family means something completely different than 50 years ago. The concepts BIRD, STARLING and STONECHAT have a completely different meaning for the children of an ornithologist than for the children of a bus driver from Colchester who is a fanatical kick boxer in his spare time. And so on.

Both complications point towards the intriguing and problematical relationship between on the one hand the fact that every single one of us meets the world in his experience and on the other hand the fact that social relationships always play a mediating role in these meetings. Philosophers who occupy themselves with response-dependent properties formulate it as follows: for every response-dependent property there are two statements that mean *exactly* the same. Or, formulated in a more abstract manner:

About the world: A thing T has a quality R.
About us: Every one of us with normal language skills who is in favourable circumstances will determine (experience and say) that the thing T has the quality R.

If the statement about the world is true, it means *exactly and nothing different than* that the statement about us is true. And conversely, if the statement about us is true, it means *exactly and nothing different than* that the statement about the world is true. By saying something about the world, you are actually saying something about us. But also, conversely, by saying something about us, you are actually saying something about the world. Have a look at these two concrete examples:

About the world: That tomato over there is red.
About us: Assuming that your knowledge of the concepts TOMATO and RED is normal and you are observing the tomato in favourable circumstances, you will determine (experience and say) that the tomato is red.
About the world: The team are playing together exceedingly well.
About us: Assuming that the hockey coach uses the concepts TEAM, EXCEEDINGLY WELL, and PLAY TOGETHER in their normal meaning and that he is observing the match in favourable circumstances, he will determine (experience and say) that the team are playing together exceedingly well.

The complications that I pointed out in this section are related to the words 'normal' and 'favourable'. How can we know that the language skills involved are *normal* and the circumstances are *favourable*? And who are the people who know this? *Who* decides what is normal and favourable? And *how* is this decided?[3]

5. Looking for trouble

The five of them are in the clubhouse: Charlotte, Emily, Sarah, Josephine and Rosemary. They've just lost another match.

— We didn't have a cat in hell's chance.

— It was awful. 7–3!

— And the way Manuel wants us to play . . . it is really hopeless. Keeping the field narrow.

— That we can't go back in defence, Sarah and me. That we need to stay on their half.

— That we can't do zone coverage. His views are really hopeless.

— And it has been like that all season.

— And all these ludicrous training methods! I really don't feel like playing hockey anymore.

— I dread going to training now. And I used to have so much fun playing hockey. He's a real moron, that Manuel.

— The way he stands there in that awful suit of his. And the way he looks . . . God, he must think he is really something.

— And all that yelling. Always complaining. Always negative.

— I'm not sure how long I'm going to stick it. Just imagine, hockey used to be everything to me. And now, this season . . . it is so awful!

— The way he stands there. He must really think he is the bee's knees. What a loser.

— Have you seen his hair, pffff.

— And those shoes.

Charlotte starts to laugh. It sounds unpleasant, bitter.

— Come on, this is hopeless. Look at us all sitting here while Manuel is long gone, straight after the match. I really don't understand. What is he doing here? Why is he involved with us anyway?

— And then going to the board, pretending to be Mr Big Shot. He really does that, you know. Going to the boardroom. Complaining. Blaming us. Oh . . . he's so good, the hockey coach from Spain. They are the best. Ha ha. Bloody loser.

— Yes, but in the meantime it is no fun for us anymore. We lose every week. We are not even allowed to think about strategy anymore. We're just supposed to listen, and do exactly what he says.

— And he doesn't know the first thing about hockey. I have no idea why they hired him in the first place. And those ridiculous training sessions.

— And the way he stands there. Such an arrogant twit!

They are silent for a moment, drinking tea and lemonade. They are staring into the distance.

— Actually we should tell him, don't you think?

— But how can we do that? Do you think he will listen to us? He's not capable of listening, he doesn't even know what it is. He's simply always right, he thinks.

— He's such an idiot.

— Perhaps we should tell them when somebody from the board is there. And with the whole team. Everybody hates him.

— Well, Sophie won't like it. She's really his favourite.

— Yes, but do you think Sophie plays that well? I don't know. Did you see the way she messed up today? Three penalty corners, and all three were her fault! Jesus . . .

— But still. Nobody else enjoys it anymore. He's messing the team up. Surely he must see that, too?

— How can he not see it?

6. Now what?

We share our world. We share our language. And we share our experience. But a great deal of divisive material is hidden in what we share. There is a great deal of misunderstanding in what we share. For example, who do I mean when I say 'we'? Does that include you? Does that include me? And how about Manuel, the hockey coach? Does it include him? Descartes? And how about that friend of mine who speaks Mandarin Chinese with a heavy Celtic accent? Do I include her when I say 'we'? And who decides this? It is so easily said: we share our world, our language, our experience. We? Our?

In our experience the world and language overlap. Our world and our language. They are entwined in our experience; they become an inextricable tangle. In order to get a foothold, you start pulling on a loose thread, start picking at a knot. 'As long as I can grasp the difference between subject and object', you think, 'or between primary and secondary qualities.' As long as it is clear to you which are the response-dependent properties and whose properties they are. As long as you know which mechanisms produce which responses. As long as you gain insight into language acquisition and know how children learn concepts. As long as you can differentiate between normal and abnormal language skills. As long as you can differentiate between favourable and unfavourable circumstances.

Intellectually we are under the spell of modern science and its obsession with objective facts. We want to be on solid ground and are looking for certainty, a secure footing, a reality that is rock solid, that gives us unambiguous and definitive answers, that leaves us no other option than to accept something that is simply a fact. An Objective Fact. The Truth. In the meantime we can accept that this will be a long road, perhaps even an endless road, but even so, we think, a road which is more or less a straight line, which brings us closer and closer to a reality that exists within itself. This is the road along which science has placed its milestones to remind us permanently of the neutral, unbiased, impartial and objective perspective that we need to respect so that we do not stray from this road.

Since we are under an intellectual spell that promises us increased knowledge, it is difficult for us to accept that the whole idea of a reality that exists within itself is an affordance of our modern, science-oriented language community. In this language community, two presuppositions secretly play the obscure leading role, presuppositions that can bind us and that can divide us. We presuppose that scientists have normal language skills, and that cognitively they are in favourable circumstances. Ironically, these presuppositions can be neither investigated scientifically nor falsified without implicitly accepting them.

This is why it may be better to take a different attitude. As long as we realise that it will remain a struggle, that we will plod on from one case to the next, and that there will also be trouble along the way. As long as we realise that much of what we share is misunderstanding. As long as we realise that the things we do not share are *still* shared by us, but in a different way.[4] We watch the same match, even though to each of us these are two completely different matches. We use the same words, even though we mean and understand completely different things. We are continuously on a different wavelength. But other people can always point this out to us. And that is the good thing about sharing misunderstanding, and about the realisation that my world is not your world, but that together these two worlds form *our* world. The same holds for experience and language. My language may not be your language, but together our two languages form our common language. My experience is not your experience and vice versa, but these two form our common experience. It is the division that we share, which is good, as it makes us both realise that we are on the wrong foot. This is not a smooth progress, it is a struggle. In a way, asking for misunderstanding is a good recipe for losing the match.

If it is about understanding and insight into our situation, into our world, our language, our experience and our behaviour – if it is about this, our common sense is of more use than our expertise. At least, if that expertise is based on the self-delusion that is part of a perspective that does not allow for distortion, a perspective that pretends not to be a perspective. Or in other words, if that perspective is built on the idea of a reality that exists within itself without insight into the fact that *that* idea is an affordance of a very specific language community: scientists who regard their language skills as normal and their cognitive circumstances as favourable.

However, if you use your common sense there is a lot to be gained. Note the entitlements and the obligations that are the result of your expectations. Do your best to regard each others' behaviour as understandable, efficient and good. Develop your trust and your accommodation, so that you can muster up the emotional stability to look at the wrong wavelength that you are on, even though it was so clear to you that it was the correct wavelength. Adopt the investigative attitude when looking at your own presuppositions about your normal language skills and your favourable cognitive circumstances. Of course, this will not lead directly to winning the match. You will not receive that certainty. Failure is part of the process. But if you are not daunted by the realisation that besides understanding you mainly share misunderstanding with other people, if you dare to accept that this misunderstanding is your fault, that your circumstances are unfavourable, that your language skills are abnormal, that your experience, language and world is not ours, then you will have found the key that you can use to gain access to *our* world and *our* language.

It is the girls in the sports bar, rather than the mirror at home, that can show you your charm and your charisma. They are your affordance for self-knowledge. You have to learn to see that. You will have to dare to see that. And you will have to be able to see, rather than charm and charisma, a loser who completely spoils things for the hockey team. Because once you can see that loser, once you can learn how to look at your own behaviour from the perspective of your supposed adversary, then there is a greater chance that you will learn to see and express completely different qualities of yourself. There is a great chance that this will make you much more charming, and that you will turn out to be valuable after all, and who knows, perhaps even charismatic.

Notes

1   See Philip Pettit's defense of global response-dependency in Pettit (1991).
2   See Hacking (1975) for a wonderful investigation into how the awareness of the role of our shared language was still absent in seventeenth century philosophy.
3   I have done some more academic work on this myself in Bransen (2001a).
4   These are huge issues that have been central to modern philosophy since the linguistic turn. I consider myself to be close to pragmatists such as Dewey (1941) and Haack (2013), but also sympathetic to a perspectivalist view such as Williams (2002), and Goodman's irrealism (1978).

9   Life satisfaction without policies

1. Social worker

His gestures are too grandiose for his small office. His voice is too loud and his words are too big.

— Look! he says.

He gets up; his chair rolls away, stopping against the wall behind them. He spreads his arms.

— Look, I can see it all from up here. I can see the boys being bored. This is where they meet, just here in the square. There used to be a supermarket below. The GPs that have taken over have just as many customers. And it all costs loads of money. People have no idea what to do. It's their neighbourhood. They don't want to leave, but they see that everything is changing and it no longer feels like their neighbourhood anymore. That is difficult for them.

— Look, it is exactly like Blaise Pascal said. I think about it a lot when I see the people here below on the square. A human being is small and trivial, an insignificant being in an enormous, incredibly large universe. There is something special about this small human being though, because he is the only one in the whole wide universe who knows of the existence of that universe, the only one who can map the immeasurable universe.

He is silent for a while, in a slightly dramatic way.

— I think about it a lot when I look at the people in the square below.

He gives me a pointed look but doesn't explain why it makes him think of Pascal. Then he goes around his desk, takes a few large steps to the other side of the office and continues in his loud voice.

— Look, those old codgers used to play here themselves. They also used to hang around here, playing ball games. Backyard cricket, sometimes using a pebble, or hopscotch. They were at St Andrew's primary school, and that was all the education they had. Starting at 14, 15 years old in the mill with their dads, or doing odd jobs for doctors or lawyers in the better neighbourhoods. They had no idea; it was just the way it went. More education wasn't part of the deal for them. That was for other people.

He is silent again.

— That is it, you see. Pascal said something else as well. A human being knows so much about the universe, and this knowledge is steadily growing. But he doesn't know much about himself. A human being is an enigma to himself. He has no real idea what triggers things in him.

The social worker looks triumphant. He walks back to his desk and sits down.

— Look, those boys have no idea. That's just it. And it is the same with the old codgers downstairs in the waiting room. They have no idea where to start either. I am trying to talk some sense into them. I take initiatives, organise an open evening now and then. I try to explain to them that the council has money for this, and that they have installed me here to see what we can do together. That there is money available.

He hits the file of papers in front of him with his flat hand.

— There are so many possibilities. I have made sure that the football club is involved, the GPs, the community support officer. The school wants to help. There is money available. And I have plenty of ideas. That should be enough, don't you think? They wouldn't know where to begin. So I just let them join in, let them be participants. That is great, isn't it?

He puts his elbows on the desk, spreads his arms and opens his hands, as if he is receiving a blessing.

— Social work is a rewarding job. They really need me here. It was a good initiative by the former government for these deprived neighbourhoods. This is a great situation for me. There's so much to do.

2. Quality of life

'All right!' The answer comes as easily as the question, 'How are you doing?' And then you just wait for what else is coming, how people fill in the idea that they are doing all right. 'Not bad', 'Busy', 'Can't complain'. You can ask additional questions or give them a meaningful look. You can also fill it in yourself, ignoring what they said, ask them about the well-being of their partner, parents or children, or raise a completely different subject, such as a topical issue, a football match or a thunderstorm that has been forecast.

Ah well, what usually happens in such chats? They are a bit like wallpaper, street furniture, or the unobtrusive knick-knacks that you have gathered throughout the years and that have all found their natural place on the mantelpiece. In fact, 'all right' is not really an answer to what actually was not really a question. These phrases are merely a number of words hanging around near us, just like a group of adolescents in a square. To some extent, they fill up our social reality: they give us a foothold in the way we deal with each other. They do not mean that much. But it is intriguing that these are exactly the words whose meaning philosophers would like to know when they explore the significance of life. Because what does it mean if someone is doing all right? When is someone doing all right? What is life all about? What do we mean by the concepts GOOD, HAPPINESS, and WELL-BEING?

In philosophy it is a well-known fact that the words that puzzle us most when we try to systematically analyse their meaning are also the words that we use correctly and without thinking in our everyday dealings with each other. 'Yes, I'm doing all right. Really. What are you looking at me like that for? And how are you doing yourself? You are looking well.'

It can mean all sorts of different things, 'all right.' Some people are happy going hiking on a Sunday afternoon, and others are happy with a good book and a cup of tea. There are people who tire themselves out at the gym for hours, people who love to go to noisy, crowded nightclubs, people who love riding a motorbike in the country, people who lock themselves away for a year to write a book, and people who thrive on improving downtrodden neighbourhoods. All these people exist and in an infinitely larger variety than I could even begin to make up. There are even people who like to go to meetings!

This evokes all kinds of questions, two of which I would like to discuss in this chapter. First of all, if happiness means something different to everybody, how can I find out what it means to me? And if we all need something else to be happy, how can we make sure that everybody is doing all right?

In the modern era, these two questions have generated the same answer: make sure that there is a powerful environment that is able to support any type of life satisfaction. Make sure there is a generously facilitating infrastructure that is good for everybody, and therefore also for you. In other words, it is not necessary to know what happiness means to you specifically. First help improve living conditions, and then happiness will find its own way.

Apparently, there is something to be said for this operationalization of the question about a good and happy life for everybody. This question is general and ambiguous, and thus it can really dumbfound us. Just try it yourself: what does happiness mean for a human being? What is well-being? What is the meaning of life? These may well be good topics for philosophers, for poets and clergymen, for at the weekend, or in the pub, or in church: for when you are having a good discussion or when you have some time for reflection. But on weekdays there is work to be done, reality is inexorably present, there is a society that needs to be led, a country that needs to be governed, a future that needs to be made controllable. On weekdays we need scientists, politicians and policymakers, people who know how to tackle things, people who ask a different type of question, questions that can be answered, operationalized questions that give concrete solutions to the grave societal problems of our times. On weekdays, the operationalization of the question about a good life works perfectly well. Even if we do not know what makes people really happy, even if we do not know what a good life looks like, there is still a lot of facilitating to be done, there is still a great deal that is good in any case, good for everybody, no matter how they envisage a good life. Let us therefore concentrate on the infrastructure. Let us make sure there are better living conditions. That is good for everybody, and sorely needed.

Still, there are two disadvantages to this modern focus on the quality of our environment. This care for our environment alienates us from the care for ourselves. Moreover, the attention to the quality of our environment reinforces our feelings of helplessness. These are the feelings I will use at the end of this chapter to make an apparently curious and counterintuitive suggestion.

3. High-quality infrastructure

In his 1971 book *A Theory of Justice*, John Rawls, undisputedly the most influential political philosopher of the twentieth century, describes an intriguing thought experiment (Rawls, 1971). Just imagine, he writes, that you are behind a veil of ignorance and that you are asked about the kind of society that you would most want to live in. Behind that veil you know nothing about your own special personal characteristics and circumstances. You do not know whether you are a man or a woman, clever or ignorant, rich or poor, in good or bad health, religious or non-religious, higher or lower class and so on. You do not know whether you like playing tennis or going fishing (or going to meetings), whether it is important to you to raise your children religiously, whether you want to take them to the theatre, or whether you would rather read a book while they are doing the monkey bars in the playground. As a result, you don't know whether you need to make sure that there are tennis courts or fish ponds, Sunday schools or theatres, libraries or playgrounds. Even if you do not know what is important in life for you personally, Rawls thinks there is still something important left for you to choose. He means choosing a societal infrastructure that benefits you whatever it is that you want personally. He means political principles that guarantee that in your society everybody's freedom will be as great as possible and the scarce commodities will be shared fairly. In other words, even if you have no idea at all about the life that you will want to lead or could lead, you will still know in what societal environment you would like to live that life. It doesn't matter what concrete shape your life takes, you will always prefer to live your life in an environment that facilitates everybody's life as optimally as possible.

Rawls used his thought experiment as an argument for his theory of justice, which revolves around the fundamental importance of a certain concept of freedom. That is not what I am interested in here. I only mention this thought experiment because it reinforces an interesting presupposition. The thought experiment only works if in the background we think that we can take good care of our own lives by taking good care of the circumstances in which we will be able to lead our lives. I will argue in this chapter that this presupposition has an unfortunate flip side. But initially it seems to be evident: of course I am looking after myself if I create an environment that will behave kindly towards me. Almost automatically one thing leads to the other. When I'm tired and feel sleepy, I take good care of myself by providing a bed. And the interesting thing is that working towards a facilitating environment will often turn out to be beneficial to everybody. If I make sure that my neighbourhood is wonderful, this is not only good for myself but also for all my neighbours. A good infrastructure is beneficial to all of us. Rawls calls such things primary goods, things that everybody will strive for, independent of all else that they would strive for.

There is an interesting domain of human existence in which providing favourable conditions seems emphatically the only proper form of care if we are concerned about the quality of life. This is the domain of upbringing, development and education, the domain of care for our children. The idea was simply and convincingly explained by Joel Feinberg in his influential article of 1980, 'The Child's Right to an Open Future.' Feinberg shows that if we are involved in our children's future plans, we are in fact behind such a Rawlsian veil of ignorance, since we do not know at all what our children will turn out to be, what they will enjoy, what they will be good at, what they will feel at home with, where they will be in the right place. We can imagine all sorts of things and may have quite outspoken ideas about this. For example, we may feel certain that our daughter has a great talent for gymnastics. We can then decide to submit our 3-year-old to a strict exercise regime. We might try to justify this by the observation that if we wait until our daughter has herself discovered whether she likes gymnastics or not, it will be too late for her to become an exceptional gymnast. Therefore it is better to take the gamble and hope that training and success will shape our daughter's preferences so that there is nothing she would rather do than gymnastics. We might also argue that generally speaking we do not wait either until children can indicate what language they would most like to speak. And there is more. We provide a varied menu so that our children will not become fussy eaters. We teach our children how to do sums and to read and write. We do all these things without asking our children for permission, so why wouldn't we try and make our daughter an exceptional gymnast if we are sure she has the talent for it?

However, Feinberg hands us a counterargument by indicating a relevant difference: a native language is a primary good, as are arithmetic, reading and writing. Whatever the child will decide to do later in life, she will want to be able to talk, do sums, read and write. These are skills that every child needs to develop, whatever she will want to do later. This does not hold for gymnastics, and this is why Feinberg argues that we should not meddle with the future of our child. We are not allowed to simply try and make a gymnast out of our child, even though we are allowed to simply immerse her in her native language. The child has a right to an open future, a future in which she retrospectively gives us permission to do the things that will be good for her in any case. We need to bring her up, form and educate her, so that whoever she turns out to be, she will be glad we did exactly the things that have turned out to be good for her.

More concretely, the idea of an open future means that we should aim our care for our children especially at creating favourable circumstances. According to Feinberg, these are circumstances that most optimally facilitate the development of the autonomous adults that our children will grow into. These adults will eventually have to find the answer themselves to the question of what makes their lives meaningful, what causes their lives to be of good quality, what they really enjoy, what really, intrinsically motivates them, and what it means to them to feel at home in their own lives. We will have to facilitate the development of their autonomy. Therefore we have no choice but to make sure that the circumstances are favourable in a general sense.

The circumstances are of completely different natures: material, social, cultural, emotional, intellectual, sporty and/or religious. In all these dimensions there is room for the individual and the personal, for concentrating care on this particular child. But especially regarding this point, the attention for circumstances is also an important reason to show restraint, to make sure you don't fill in too much, to avoid being too indulgent. After all, your child doesn't really know yet what she wants, what she really wants. Just like her parents, your child does not know in a specific, personal sense what she, later, as an adult, will want her younger self to have wanted. That is exactly the meaning of the veil of ignorance; with a view to the child's right to an open future, we should pay special attention to favourable circumstances. We shouldn't make it too personal but should provide conditions that facilitate the child in her development, in whatever specific direction she will turn out to develop. We will provide food, shelter, education, a safe bonding environment, intellectual inspiration, challenges, comfort and so on.

In any case, that is the presupposition that is reinforced in the current style of thinking about happiness, well-being and the meaning of life. You take a minimum of personal risk and contribute to the happiness of the maximum number of people by concentrating on the quality of your environment.

4. A knotty situation

She hadn't noticed before how much turbulence Nigel creates. She thinks he is a nice guy; it's not that. He's always willing to help. You can go to him with even the smallest problem and he will always take you seriously and will always try to help. It is really amazing how much energy he has and how much this neighbourhood means to him. That was clear this evening as well. She had arrived a bit early to help with the preparations, like making coffee or setting up the room, but it felt as if she was getting in his way. He was already working on the coffee, the plastic cups were on the tables, he was busy straightening the chairs with his mobile jammed under his chin. He was picking up piles of four or five chairs from storage, carrying them along the corridor, all the while shouting loudly into his mobile that the community support officer should get a bloody move on because he didn't want to start late and that all the organisations should be there before their public arrived.

— Miss Irene is already here. At least she is reliable.

She had to smile at his use of 'Miss'. When she walked to the table at the front of the room to see if all the brochures were there, he put down his pile of chairs, abruptly ended his conversation, and walked straight over to her.

— Look, he began. Good printing, don't you think? Picked it up from the Print Shop this afternoon. That's another thing: they can only ever have things ready right at the last moment. If you knew how often I'd rung them . . . And you wouldn't believe how many spelling errors I had to correct. Nobody seems to learn how to spell at school anymore. This was written by an architect, too. Unbelievable.

Then he put his hand on the pile of brochures. It was as if she wasn't allowed to look at the brochures, which was odd because it also seemed like he wanted to show them to her. He didn't seem to notice himself, and turned around swiftly leaving the pile of chairs where he'd left them, as if he had suddenly remembered something else. He walked to the sound system, muttering to himself.

— You have to do everything yourself in this place. Just look how these cables are all tangled.

He started pulling randomly at the cables and then she suddenly realised: she could see he was making it worse, and that he was making the knots tighter without really paying attention to what he was doing.

— Shall I put the chairs out then?

She shouldn't have said anything. While holding on to the cables, Nigel turned around to the pile of chairs that he has just abandoned. The cables tightened and he pulled over the two microphone stands, causing one to knock down the jug of water that was ready on the table for the speakers. He swore loudly, glaring at her, and then apparently realised what he was doing. She couldn't help it, but the perplexed expression on his face made her burst out laughing.

5. Self-alienation

The country is doing well; in fact, the country is doing exceedingly well. The shops are full, the trains run on time, the government can be contacted online, and there is an awful lot of money: to save the Greek economy, to go on holiday six times a year, to finance university super-computers, to continuously surf the Internet on your mobile, to house schools and museums in ornate buildings, to widen the motorways throughout the country, to hire social workers and to give them a big bag of money to spend. The country has been completed: the quality of our infrastructure is relentlessly good!

But despite our stable government we are massively dissatisfied and disappointed with politics. And research tells us that as many as one in twenty of our fellow citizens is on antidepressants, and sometimes has been for years. By the way, that is another thing that has been excellently organised: all these pills find their way to the patients, through doctor's prescriptions, well-furnished pharmacies and generously contributing health insurance companies. So what is the matter? How is it possible that so many people are apparently doing so badly in a country that has been so successful at organising a high-quality, facilitating infrastructure?

I will make a suggestion, no more than a suggestion. I do not have a great deal of supporting evidence, but that is not what I am interested in here. I want to make you think, and I hope that my suggestion is sufficiently intriguing. Perhaps one thing has to do with another. Perhaps the explicit focus on improving the quality of our facilitative living environment negatively affects our insight into what people really need in their lives. Perhaps this dominant attention to living conditions has not only created a well-oiled infrastructure but also a blind spot regarding life itself. Perhaps we have lost sight of what life is, of what humaning is, the verb, because we have focused all our attention on the conditions. My suggestion is an echo of the complaint that you sometimes hear about education, that it seems that its only value lies in how the skills learnt can be applied later. As if education in itself is only a precondition. As if, in the words of Alan Watts, the only purpose of a sonata is its final chord (Watts, 1951, p. 116). This leads to a remarkable type of self-alienation. If we only pay attention to the conditions necessary for a good life, we may well lose sight of that life itself.

Our extraordinary plasticity, our exceptional adaptive skill, is not only our strength but – as always – also our weakness. As a specimen of *Homo sapiens*, we have started to cultivate our environment but probably also ourselves. Perhaps this is why we cannot find our real needs among all these other needs that we are supposed to have, that are prompted. Perhaps they no longer seem to be worth our while, our real needs, because it is too tempting – and too easy – to lean back in the soft cushions of our new lounge sofa. Because really: what can we actually know about what really drives us, what really makes us happy; what can we really know about our own well-being?

Arnold Gehlen has developed an interesting view of the problems related to our insight into what really motivates us. Gehlen states that the distinction between actual needs and prompted needs is misleading. There are no original needs that have been pushed out by cultivated needs. There are only cultivated needs. For us, these are of vital importance, according to Gehlen, as they are linked to our cultivated goods, and as a result these needs can be satisfied in theory (Gehlen, 1957).

Let me give you an extreme example. Let's assume that in the depth of my thoughts I am convinced that I am really an old Germanic tribesman, one of the original inhabitants of the Rhine-Meuse Delta, devoted to my region, and happiest when trekking through the floodplains of nearby rivers. Of course I know full well that it is now 2016 and that I live in a house built in the 1970s, somewhere in a village near a river, that I cycle to work through the meadows and write philosophical texts. Still, at the same time I realise that I am really happy and feel completely at home in the immeasurable emptiness of the landscape, with all that water engulfing the floodplains. What should I do with my life, assuming that the whole idea is plausible and I am indeed a Germanic tribesman at heart? Should I resist the government's plans to raise the river banks? Should I dislike the pragmatic way people in this country think about rivers and dikes? Should I lock myself away in a reservation, together with Konik horses and Galloway cattle?

Such silly questions, Gehlen would think. Asking them clearly betrays the simplicity of my thinking, as if I think there could be something like a real, actual, original, authentic desire. Why would the desires of the Germanic tribesman in me be more original than my actual desire for a cup of coffee from the machine down the hall? My tribal desires are cultivated desires, too. And you can see the sense in them, Gehlen would emphasise, since it is completely suitable for an old Germanic tribesman to desire a hike through the floodplains, as this is almost the only thing his environment facilitates for him. Well, this is the crux, Gehlen asserts. I long for a cup of coffee simply *because* there is a machine down the hall, just as I can long for a new app for my mobile because it is available, and just as I can long for a hike with two walking sticks because Nordic walking has been introduced to this country.

This is what Gehlen is really interested in. The only desires that we could and should care about are cultivated desires, desires that have become ours as a result of the culture that we live in. If we start talking about authentic desires, we forget that even our deepest motivations are only about coping, about surviving together as a species. This also means – and this is of crucial importance to Gehlen – that it should be possible to satisfy our needs. And they can be satisfied exactly because of their suitability in the culture that we live in. In Gehlen's view, looking for real, original desires outside of this culture is looking for the dark depths of ultimate primitivism, where we will not be able to find anything but arbitrary and subjective feelings. A disastrous route, according to Gehlen.

I am not interested in whether Gehlen is right or not. I can't really imagine what this could mean at this point. But he does point out the sore spot: the problematic concept of authentic desire. He takes a clear idea to its extreme but logical conclusion: don't think about your own, intrinsic, original, authentic desires, but just be one of us. Acquire those desires that are likely to be satisfied in this culture. Concentrate on your environment, on the infrastructure that facilitates your life, and adapt yourself to what this environment, our culture, has to offer you. Forget your originality, your individuality, your uniqueness. It doesn't matter anymore. It doesn't give you anything tangible to hold onto. It can be no more than a primitive whim, as useless as the angry and helpless screams of an exhausted toddler, throwing himself dramatically to the floor because he wants a bag of sweets.

In my view, Gehlen has taken his conclusion a bit too far. On the one hand, it must be admitted that he has a point: we have no idea what we should do to discover what well-being and happiness really mean to a person. But that doesn't detract from the fact that you really miss something if you focus completely on your environment and provide a high-quality infrastructure that is satisfactory in every conceivable sense (material, social, cultural, emotional, intellectual, sporting, religious and so on). Because then you end up with the realisation that this environment can give you anything your heart desires . . . but what is that: what is it that your heart desires?

Have a good look at that question for a while, I would suggest. Let it sink in. Do not ignore it. Do not jump – like Gehlen – to an answer that denies the reality of this question. Not knowing what you want, and *knowing that you don't know what you want*, is an important form of self-knowledge. Thinking that well-being and happiness are only a matter of favourable living conditions, of high-quality infrastructure, of a facilitating environment, is opting for a special type of self-alienation. We sometimes come across this type of self-alienation in stories that we have all heard before, stories of people who seem to have everything their heart desires: a beautiful house, lovely children, healthy parents, a great job, a loving, beautiful, caring and also still sexy partner, and yet . . . And yet.

6. Helpless

There is another problem with the idea that the best way to achieve a successful and happy life is to provide favourable living conditions, as this idea gives each one of us a feeling of helplessness. In a way, it makes us feel like the mirror image of free riders. I will try and explain what I mean. The train can only run if everybody pays a share in the costs and so everybody buys a ticket. Everybody? Well, if nearly everybody does so, you can be a free rider and take advantage of this. The train is running anyway. Together we have made sure that the infrastructure is there, and if I were to forsake my duties as a citizen and travel without a ticket, nobody would notice. The disadvantage to us all is infinitesimally small, whereas the advantage for me can be rather large. Not a good idea, of course. But what if we consider the opposite situation? Suppose the streets are full of rubbish: litter everywhere, half-eaten rubbish bags galore, dog poo on the pavement, old furniture, you name it. And there you come walking along, having just finished an apple or a healthy muesli bar. You look for a bin, find one, see that it is overflowing and notice all the garbage around it on the ground. What do you do? Do you keep your head held high and try to cram your apple core into the bin? That is a good idea of course, but would it help? Will you set the right example? Will you be the laughing stock of the neighbourhood?

Criminologists have come up with a good phrase to explain the behaviour that you are most likely to display: the *broken windows effect*. Nobody wants to be a loser in the eyes of other people and so we effortlessly adapt our behaviour to what seems to be the standard. If it's an awful mess, then we become a little bit messier ourselves. So I wouldn't be surprised if you were to launch that apple core into the bushes (I am not so sure about the muesli bar wrapper. I know how unthinkable certain behaviour can be to eco-conscious people, especially the first time. But perhaps you can do it surreptitiously, let it glide from your hands as if by accident.). The broken windows effect also works the other way around: if everybody is friendly, caring and decent, we will try to emulate this positive behaviour. However, this means that everybody will realise that there is not much you can do about your own circumstances. We can do something as a species, and also as a group. Governments can do it. But individuals, concrete people, cannot do much about their living conditions. They can take part, they can let themselves be inspired by a social worker or by a charismatic leader. We can do a great deal when we are on the crest of a wave of a popular movement, each one of us, but exactly that also reinforces the feeling that we are actually, each person individually, powerless when trying to create a facilitating environment. The Arab Spring caused people to be over the moon for a while, but then we had to go back to the grindstone: there was work to be done. If we have lost the old, hostile and harmful environment, then it is time to create an environment that really is facilitating. And this is exactly the moment when nobody seems to be in control and when people look helplessly and powerlessly at how the old infrastructure has kept its persistency and how it manages to insinuate itself into the fancy talk of progressive politicians. Gehlen could not have come up with a better scenario.'

'You must be the change you wish to see in the world.' This may sound good at first, but implicitly it only emphasises the helplessness of the window *makers*, or, in other words, of the mirror image of free riders. And it does so precisely and subtly by the unthinking way the words have been used, reinforcing without noticing the presupposition that for a good, successful and happy life you need favourable living conditions, a facilitating infrastructure. Have a good look at the words used in the saying: 'change' it says and 'world'. So it is not about changing yourself, but about improving the world, our living environment. And even if you have improved yourself, it has only just begun. Then the rest needs to be taken on. Can you imagine yourself, next to the bin, with the healthy muesli bar wrapper in your hand? I get tired only thinking about it. Why would I have to change the whole world? Why the focus on our environment? And why would it have to be *me*? Have you any idea how difficult it is to understand oneself? It will take you at least a lifetime. And then we would like a second go, so that we can do it a little bit better next time round.

'You must be the change you wish to see in the world.' It doesn't bear thinking about . . .

7. Now what?

Something interesting has happened to us, behind our veil of ignorance. Because when we are there, there are two things that we do not know: we are ignorant not only about our specific needs and desires, but also about the conditions in which we can and must live. And then something special happens in Rawls' thought experiment, something that must make us think: because almost without thinking the situation behind the veil is presented as a situation in which you will have to choose the circumstances in which to live your life, a life of which you do not know what it will be about, what happiness, well-being and sense will mean in that life. But why would that be the choice you need to make behind the veil? Why will your happiness depend on the societal circumstances that you choose? Why couldn't you look at it in a different way? What I mean is the following. You find yourself once again behind the veil of ignorance, but this time you're asked what desires and needs would be best for you to choose, given that you will have to live your life in unknown circumstances. Suppose that this is your question, behind the veil. The situation remains exactly the same, but you shift your attention and now salvation is no longer expected to be the result of societal circumstances but of your own character. If you imagine that your fate will depend on this, that this is the choice that you will have to make, it seems evident to me that the best thing to choose for are very flexible needs and desires, in other words a huge talent for contentment. Because wherever you end up, if you are good at adapting to the conditions that an environment has to offer, you can be happy anywhere. It doesn't matter then in what society you eventually find yourself. To be honest, I think that a talent for contentment would be the best choice for everybody who finds themselves behind the veil of ignorance.

To me such a talent seems like a wonderful ability. Especially if you can choose your own character behind that veil of ignorance where you do not know whether you are a man or a woman, clever or ignorant, rich or poor, in good or bad health and so on. Such a talent will certainly be helpful if you do not know whether you like playing tennis in a country without tennis courts, or fishing in a country without water, or going to meetings in a country without a culture of dialogue. In such conditions, the ability to reconcile yourself with your fate can contribute enormously to your happiness.

By the way, I can imagine that you have grown so used to all this obvious concern about our living conditions, that my plea for contentment is slightly problematic for you. Are you wondering about my social engagement? Are you perhaps thinking that I am naive? Do you suspect that I am in fact presenting reactionary advice? Do you think that I am closing my eyes to the shrewdness that the rich and powerful will undoubtedly employ to benefit from people who have been encouraged to reconcile themselves to their fate? But what about you then? Could this unease and suspicion against the plea for contentment be the result of a blind spot? Have you ever wondered what it is that you might be closing your eyes to?

In this chapter I have tried to raise a question rather than give an answer. If I have made a case in this chapter at all, then it was a case for an investigative attitude, because I do not know exactly what I mean when I say 'I have a good life'. I don't know exactly what I am suggesting when I discuss the quality of my life. I am also not so sure whether the quality of my life is dependent on the quality of my living conditions. That is why I have no idea whether the quality of the infrastructure of this country actually contributes in a positive or negative way to the quality of my life. That is why I have drawn the following conclusion: the question of the quality of our living conditions is an overrated and misleading affair. Perhaps they are not so important at all, these conditions, as they only play a small role in our happiness. Perhaps social work is only beneficial to the social worker. At least, as long as he is content with his job, as long as he likes helping other people – not so that they progress, but simply because helping people is something that he likes doing.

Interesting things happen when you are in a crowd of people on the platform and the trains are not running. Of course, there are always a few people who start grumbling, looking for the people responsible who need to have their ears boxed. But even they change their tune after a while and join the others who are trying to make the best of it, who are managing to create a pleasant conviviality, just being together, all strangers and unknown to each other, reconciling themselves to their fate and enjoying each other's company. These are beautiful moments, when the infrastructure fails and people need to rely on each other. Perhaps it is exactly such moments, but then spread out over dozens of years, that can be seen in the hopeful faces of people in third world countries who have once again seen their crops wither in the field, or who are trying to recover in what used to be their house after an earthquake or a hurricane. Don't get me wrong: of course I would like everybody to have full shops and trains that run on time, but I would also like everybody to have the ability to be cheerful, happy and full of life even in wretched conditions.

Is there something we can learn from this if we are concerned about our children's open future? We provide favourable conditions, so that they can be whatever they want to be. We provide warm housing, healthy and varied food, good schools. We teach them hygiene, reading, writing and arithmetic. That is what they need for any life they would like to lead. But shouldn't we also teach them to be content, as this seems to be fundamentally important for an open future?

And contentment should not be interpreted as a matter of self-control, because that is not the issue at all. Postponing the gratification of their needs is something that we have been teaching our children for ages. This is something that they do every day at school, even if they are bored, so that they can go to the appropriate form of secondary education, so that they can land themselves a good job, so that . . . This has nothing to do with contentment; rather, it only stirs up discontent. Contentment is not about later but about now. And it is not about the conditions, but about our character, our attitude, our mentality.

Quality of life is a vague phrase, even though we all want to promote it. It is a phrase that we obviously understand, without thinking, and that refers to the quality of our living conditions. In this chapter I have tried to make the case that this is in fact rather curious and dubious. My conclusion is therefore that the concept QUALITY OF LIFE requires an investigative attitude. We understand only a little of that concept, and as a result we are not nearly ready to ask concrete, direct, operationalized questions about it. It will be a while before quality of life is a question that can be passed over to science, a question that is ready for the missing information subroutine.

10 Responsibility without expert witnesses

1. In prison

She has some paint on her hands. He sees it while he is listening to her. It must be from the day-care. Her hands lie on the table in front of her, like dead birds.

— The worst thing is the regrets, that there is nothing that can be done about it. That it has actually happened.

She is silent.

— But I'm glad that he's dead. I don't regret that. Not at all. But that I . . . and that for such a long time . . .

She looks at him. He can see the doubt in her eyes. Horror. She looks straight through him. She sees him, but she is looking at something else entirely. A memory.

— It had to happen, you see? It was a dead-end street. I had been made to feel so alone, so afraid, so helpless, so broken. I was there and . . . Jesus, that I took the knife from the drawer. Oh, I knew very well that the knife was there. That is something that I had realised a long time ago. For years, every time when I used it to cook dinner. I knew that it was there. I was afraid of that knife. I used to wash it immediately, always, after I'd cut the meat. Wash it, dry it, and put it back in the drawer straightaway. And then I was thinking . . . then I was thinking of the knife. That it was there. That was good. And it felt bad. It was completely wrong. The idea made me feel calmer. And it also frightened me. God, I was so frightened. For all those years.

She sits, without moving. There is a tremor in her voice. Her eyes fill up. She takes a breath, forcing herself to breathe calmly.

— And that you *know*. While you . . . while you're stabbing. And before. You just know. That there is no way back. It all happens so quickly, but still you know. I knew. I could follow exactly what was going on. I knew what I was doing. To the kitchen. Jesus, I was running to the kitchen, towards the knife. And then back to him. He was sitting there with that stupid grin on his face. He had no idea what I was doing. What I was about to do. It would never have crossed his mind. He just thought he could do what he liked. All those beatings. All that violence. All that humiliation. It was all there, at once, it all came together, while I was there, in our living room, him on the settee, and I, while I . . . with the knife . . .

It is blue paint. Blue with a little bit of grey. He sees it. He realises that he is looking at her hands. Her hands, still resting on the table, like dead birds. As if they are not hers.

He looks at her. Her eyes are clear now. She is bridging an enormous distance when she goes on.

— And he just let it happen. With that stupid grin on his face. As if he was dead drunk, plastered. But he wasn't. He just didn't understand. It just couldn't get through to him. The knife went in so easily. Yes, that is giving me nightmares, in here.

She swallows.

— But then I was so calm. Really bizarre. Three times I stabbed him. The knife just slid in. It was really creepy how easy that was. And then I just left it in, after the third time. I went to the bathroom and looked in the mirror. Only then did I see my face, how it had been battered. I couldn't even feel it, but my left eye was completely closed. It was red and purple, swollen, and blood was seeping from my mouth. I saw that my jaw looked wonky. It was looking odd, but I couldn't feel anything. I hadn't even noticed that it was broken.

Suddenly she moves her hands. She feels her jaw. She looks at him.

— It's still wonky, isn't it?

2. Who is to blame?

Even if only something minor goes wrong, the whole country is up in arms and everybody asks who is to blame. Who is responsible for this? Who has done this? Couldn't this have been prevented? Which authority has been sleeping on the job?

It is an understandable reaction: it is in character for us, *zoon logikon* in the twenty-first century, in character with the things we say about each other and about ourselves, with our common language, the language in which we live, in which we do our humaning. Even when you're still a baby crawling around, you already learn to stay away from the stairs, not to touch the stereo, not to touch the oven or your father's coffee cup. It becomes clearer when in your terrible twos you start noticing that it is really inappropriate to lie on the supermarket floor screaming your head off because you want crisps, biscuits or sweets. You gain even more insight into your little world when you learn how to extend your power over other people's behaviour. You notice that you are welcome in the world of social interaction: using your brightest smile you can coax your grandfather into reading your favourite book to you once more. And when your aunt allows you to play on her new smartphone because of that same smile, you know that you have it figured out. But once you've reached that point, you will also have noticed that it is not one-way traffic: other people can control your behaviour too and make you reluctantly do things that you would rather not do.

It is a beautiful mechanism, the language of blame and praise, of responsibility, of valid and invalid excuses.[1] It is a mechanism that is deeply rooted in our common sense. But it is also a mechanism which you can easily lose your grip of, and a mechanism that eludes your control if you are dealing with supra-personal organisations, with anonymous authorities, with regimes, with bureaucracy and with public opinion in a complex welfare state like ours. I'll explain what I mean.

People do not only learn to control their own behaviour, but also that of other people. You can use violence to do this. This is what a robber does when he snarls 'your money or your life', as if he is giving you a choice. This is a false and perverse way of presenting things, because you don't have much choice when you are looking into the barrel of a gun. But you don't have to use force; you don't have to exert your influence in a false or perverse way. Controlling other people's behaviour is not a matter of selfishness or a lack of respect or sympathy. Whenever people work together, there is mutual control of each other's behaviour. 'Control' may sound a bit unfriendly; it sounds better if we say that when people work together, they take each other's wishes into account and they can count on one another. However, in my eyes this implies giving a little bit too much responsibility to the other person. As if you, being nice and friendly, simply expect that you can count on the other person because he will take your wishes into account. That is not the way it works, because you, with your nice and friendly expectations, are in fact influencing, controlling and manipulating the other person's behaviour. Whether you want to or not.

This is just the way it works when people have expectations of each other's behaviour. These expectations cause the other person to be curtailed in his behaviour, and his behaviour becomes an aspect of your expectations, either implicitly or explicitly. Our expectations always exert influence. Even an adolescent who wants to escape the norms and values of his bourgeois background is restricted in his behaviour and has no choice but to take into account the influence exerted by social expectations. There is nothing wrong with this; it is just a question of common sense.

After all, we also have our own expectations of other people's behaviour. Together these mutual expectations make cooperation possible and give us control over our own behaviour and that of others.

Essentially, mutual expectations lead to behaviour that takes place in an obvious and unobtrusive way – in the background. After all, what we expect doesn't need our attention. Our autopilot not only allows experienced drivers to drive their car with the greatest ease, but also takes on a lot of our work in social interactions. In the supermarket we never bump into other people with our trolley, and in the unlikely event that this does happen, we politely apologise. We hold open the door of the fridge if we see that somebody else also wants to reach in to get something, do not cut off our fellow customers to reach the checkout first, patiently wait our turn in the queue and politely answer all the questions that the checkout assistants are required to ask: 'No thank you, I don't have a Clubcard, and no thank you, I don't wish to save up for a Fun Day Out Voucher.'

In Part 1 I described this practice of mutual expectations as a fundamental characteristic of our common sense. Expectations that are based on our folk psychology have a normative character that can be explained in terms of entitlements and obligations. It is in this normative framework that the language of blame and praise, the language of responsibility, has its roots. You can recognise this language in the way we use apologies, which are an important part of this framework. You know what the other person expects of you, and that is why you know when you should apologise. You also know which apologies are valid and which are not. Of course, this is not a rule set in stone, and neither is it knowledge that is explicitly written down somewhere. Again, it is simply a question of common sense. Of course, there are related grey areas in which we need to find our way in a probing and creative way, using our investigative attitude. But besides grey, there is also a great deal of clear black and white. As a 7-year-old child, there is no longer any excuse for wetting yourself if an advert on TV is really hilarious. As a 14-year-old schoolboy, 'the level crossing barrier was down' is not a valid excuse if you are late for school for the fourth time this week. As a 20-year-old student, having just received a funny message on your mobile is not a good excuse for talking loudly during a lecture. And as a 40-year-old man, wanting to make a stand for women's emancipation is not a good excuse for refusing to hold open the door for a woman.

These are situations in which mutual expectations are clear and in which you can easily draw up a list of valid excuses in black-and-white; they are situations in which there is no problem with the language of responsibility. If people do not satisfy our expectations, without thinking we give them the responsibility to come up with a valid excuse. Similarly, if we do not satisfy their expectations, without thinking we take the responsibility upon ourselves to come up with a valid excuse. And there are even more responsibilities. We automatically give other people the responsibility to meet our expectations, just like we ourselves take on the responsibility to meet their expectations. There is more: we automatically give other people the entitlement to ask for an apology if they notice that we have not met their expectations, just like we ourselves automatically feel entitled to ask for an apology if the other person does not meet our expectations.

To a great extent, this language game of giving and taking responsibility is a social lubricant; here, our common sense shows its normative side most clearly. We want to and have to take each other to task about the quality of our humaning, so that we can do our own humaning adequately. We want to and have to build on the understandability, efficiency and goodness of each other's behaviour, and this is why we take it for granted that we can request and receive further explanations if we cannot follow the other person's train of thought or if we feel that we cannot allow this person to get away with what he's done.

This everyday practice functions perfectly in small-scale contexts in which people have, get and take the opportunity to explore the grey areas together. For example, this is what happens in the domestic circle, in friendships, in small companies and so on. In such small-scale situations – disregarding the larger-scale societal context for a moment – there is nothing wrong with striving for consensus. It is a matter of course that when you and others find yourselves in a grey area, you explore the grey area together, using your common sense.

Of course this sometimes goes wrong, too. Sometimes you feel powerless, afraid, lonely and confused if you do not understand the other person's expectations, if you find yourself in a grey fog that the other person fails to acknowledge. Sometimes it is impossible to identify any common sense. Sometimes you are certain that you are right and he is wrong, while at the same time he is certain that he is right and you are wrong. Then it is difficult to find any common ground for trust. Then it is time to start practising accommodation. Then you need to hope for the sudden realisation that you are not in your scenario and he is not in his scenario, but that you together are in a third scenario, a scenario that neither of you know, a scenario in which both of you sorely need your common sense, a scenario that first and foremost requires an investigative attitude. As soon as you both realise this, as soon as you both see that you have come to a grey area, that your mutual expectations do not match and that as a result both of you are restricted in your actions, new possibilities will arise. You will both receive and take the opportunity to express your expectations and to explain them. And then together you will look for ways to better attune your mutual expectations.

That is how simple it is in everyday practice. You may think that it is abundantly clear that you really don't feel like a romantic evening with your husband because you always go out bowling with your girlfriends on a Friday night. Your partner, however, may think that it is clear that bowling is out of the question today as today is your 7-year wedding anniversary. But if he turns out to be mistaken about the date (not until next month!), your sense of humour should be able to help you rise above your hurt feelings.

In our current society, however, it has become extremely difficult to deal with these grey areas in a prudent way. Increasingly, the public domain has become so complex, so multi-faceted, so global and so anonymous that nobody has a clear idea anymore of where and when you could take someone to task about his implicit expectations. In these scenarios of anonymity, the mechanism of blame and praise – in other words the language of responsibility – has completely gone adrift. Moreover, the mass media are really spurring us on, reinforcing the idea that somewhere there are always professionals who should have known better, who should have taken their responsibility or who should have used their expertise. Authorities, both high and low, are at a loss as to how to deal with their anonymous responsibility and are increasingly looking for support in laws and regulations. This is also difficult for the judiciary, especially the conversion from the formal language of the law to the concrete behaviour of human beings. Fortunately, you may well think, we still have science to help us out . . . But especially in the language of responsibility, in the practice of valid and invalid excuses, there is not much to be gained from science. At least, this is what I contend in this chapter.

3. Not accountable

If you take a rather silly situation as your starting point, it is often easier to see when things become interesting. You can't expect a bottle of champagne to come to you when you beckon it. And it won't help if you say please and thank you, if you talk to it in French, or if you shout at it really loudly. A bottle of champagne just does not participate in our social interaction. On the other hand, if you want to stop a maniac with a big knife from chasing you, it doesn't matter whether you whisper or curse, whether you speak in French or Russian, or whether you only gesture that you want him to stop. Threatening people with knives is just not part of our social interaction. Suppose you're at the airport and a Chinese-looking toddler grabs hold of your leg in the queue at passport control. Whatever you try, she will not let go. There's no adult in sight who the child might belong to. The little girl does not appear to understand what you are saying, and you have no idea what language she speaks. The customs officer looks at you and asks you for the child's passport. You cannot seem to make contact with the girl. At home, her parents will probably take her to ask about her difficult behaviour, but here, in this situation? You have no idea how to approach it. The customs officer is addressing you, as if you are responsible for the child. But of course this is not the case. If only you could explain the situation to him and he believed you, then it would be up to him to take the child off your hands. Would he know how to deal with her? He has probably got more means at his disposal, but he still wouldn't get far, perhaps about as far as you got with your bottle of champagne.

Here is another scenario: a young man gets on a deserted underground train. A drug addict, you think. Stoned, you suspect. East European? He approaches you and then asks you in perfect English, without slurring his words, if he can sit next to you. Before you can respond, he sits down and puts his head on your shoulder, and sighs deeply. You think that he has immediately dropped off to sleep, but he starts talking in a clear voice. An American named Stanley Milgram has given him an assignment: he has convinced the young man that you will not mind at all if he asks whether it is all right if he sits next to you. 'Cool guy, this Milgram,' the young man continues, 'because he is right.' Then he is silent and it now looks like he is really falling asleep. How can you address his behaviour?

There is a purpose to my digression and these unusual examples. I would like you to seriously consider the many advanced presuppositions embedded in even the most mundane forms of social interaction. All these presuppositions play their part, every time you take someone to task about their behaviour: the fact that this person is accountable, just like you; that he knows what it means to meet somebody's expectations; that he knows when an apology is required; that he knows what kind of remarks are genuine apologies and what remarks are not. Even if you are dealing exclusively with strangers – like in Part 1 on the fictional island of Endoxa – your folk psychology is functioning and you are able to take part in new and unknown scenarios. Here you presuppose a great deal in the background about what makes behaviour understandable, efficient and good.

In modern democracies, the rule of law has provided a safety net under these presuppositions. It is all quite neatly, unambiguously and explicitly articulated in the many different law books. Of course that is a great thing.

The rule of law is one of humanity's most wonderful achievements. And of course the rule of law is important in our current anonymous and multi-faceted society. There are some nasty edges to it, such as the habit that has blown over from the US to use the judiciary to settle scores regarding frustrated expectations. But apart from that, it is of course fantastic that we can approach strangers in an unknown environment with great inner peace and full of confidence. The rule of law gives us a great deal of guidance in such situations.

However, the rule of law has also brought about a whole new way of exploring grey areas. We don't do that together anymore, fumbling along, with sympathy, respect and common sense. It has become much too complicated for that. Nowadays, we leave the exploration and mapping of grey areas to the experts; the legal scholars, who use fancy words to write exactly and accurately which grounds for exoneration should be recognised by law. And if it becomes really tricky – for example if it is about 'diminished responsibility', which is the legal term for 'accountability' – they use expert witnesses, such as behavioural scientists and psychiatrists, to give them advice. These people are given the thankless job of objectively determining whether somebody was accountable at the moment that they committed the crime. Of course, psychiatrists have long since understood that little communication is possible with either a champagne bottle or a Chinese-looking toddler. The young man who Stanley Milgram sent to you in the underground will probably be more difficult for the expert witness, but he will certainly find a way of dealing with him, too. And to be frank, this is not such a serious case. Usually such experts have to deal with out-of-control maniacs carrying large knives. Are such people accountable at the time they commit the crime? Is such a maniac responsible for his deeds?

How does a psychiatrist determine whether somebody is accountable? Let's first have a look at the presuppositions involved in a psychiatrist's work. First of all, he will have to break down the relational attribute 'accountability' into different components. He will have to attribute some of these components to the suspect and other components to the circumstances. Then he will need to determine that the decisive component (legal competence?) is a non-relational characteristic of this suspect, or at least a property that was relevant when he committed the crime, whereas the properties of the environment were irrelevant. Finally, the psychiatrist needs to determine all this while being an impartial, neutral, objective observer, from behind a two-way mirror, as it were.

Let's consider these presuppositions one by one.

The interesting properties of human beings, but also of things, scenarios, circumstances and so on, are often relational in nature. They are properties that only count in relation to something else. An example is 'weighing less than Jan'. This is a property that my neighbour has: she weighs less than me. But if I really tried to lose weight drastically, my neighbour might well lose this property. In return she will then acquire a new property: 'being heavier than Jan'. The interesting thing of such relational properties is that you can never unconditionally apply them to yourself alone. You can lose such a property or you can obtain it, even though you yourself don't change at all. If a scientist were to isolate my neighbour for a certain period of time and carefully make sure that she remained the same weight – not a gram more or less – she could still change from 'lighter than Jan' to 'heavier than Jan' if I successfully followed my diet in that same period. From a metaphysical point of view, such relational properties are absolutely amazing; from a scientific point of view they are absolutely unreliable. If you have amassed a great deal of money in your life, which you keep in a sock under your mattress, it doesn't bear thinking about that suddenly something might happen in the world that will cause your fortune to lose all its value. Such a situation should not be allowed! That is not the kind of universe we want to live in.

You might think that there is something that could be done about this. In the olden days greengrocers used counterweights on their scales so that 2 pounds of runner beans would actually weigh the same as the counterweight of 2 pounds. Of course a corrupt greengrocer could use counterweights that weighed less than what they were supposed to weigh. But nowadays we are way beyond such a primitive stage. When my neighbour wants to weigh herself she doesn't use the type of scales where she needs me as a counterweight. 'The same weight as Jan' has long been replaced by '11.5 stone' ☺. This still remains a relational property in one way or another, even though scientists now have extremely stable and accurate calibration methods at their disposal.

It seems like a good procedure: replace relational properties by properties that are in fact non-relational because they can be expressed in objective, unchangeable, absolute units. But that puts you right in the middle of the problems regarding the response-dependent concepts that I discussed in Chapter 8. Just have a look at an everyday property: 'irritating.' Suppose that as a teacher you have a child in your class who is continually irritating. Suppose you say to him, 'Stop being so irritating, James!' And suppose James is not only a troublesome but also a clever pupil who quasi-politely corrects you: 'Well, no, I don't really think I am irritating. But apparently I am irritating you and perhaps that says more about you than about me.' How are you going to approach a situation like this? And of course there are more response-dependent concepts than only 'irritating': amusing, hideous, brave, friendly, intelligent, careful, decisive, caring, strong, sincere, deceptive, evil, unreasonable, to name but a few. The list of human properties that are response-dependent is virtually endless. And yes, 'accountable' is also on that list.

Of course, you can try to break down the response-dependent properties into different components, and attribute one component to the object and another component to the subject. In the local context of your classroom, this is clearly part of the interaction. Your pupil irritates you and this is caused partly by the expectations that you have of him and partly by his behaviour. Of course, as a good teacher, you do not enter into a discussion about your expectations, but you simply and confidently presuppose that your expectations are normal and that James's behaviour therefore is irritating. We all know that such situations are not always as clear-cut as I have just presented, but in the black-and-white areas of our daily communication in the classroom this is never a serious point. As a teacher you are simply right.

Only in grey areas does it become tricky, and interestingly enough it is in these areas that the grey itself is also a response-dependent property that teachers can keep under control in all sorts of ways. Dealing with grey is a delicate affair, as I have argued in Chapter 5, and nowadays it is even more difficult for teachers, as your pupils now enter the classroom bearing a scientifically sound diagnosis. For example, James may have the letters ADHD stamped on his forehead and you are kindly requested to remind him to take his medication even though you would rather be putting him in his place. In the personal and intimate atmosphere of your classroom, this is something that you will have to sort out together with James, because you will probably have some warm feelings for him anyway; otherwise, why would you be working at a secondary school? The diagnoses based on the DSM simply provide extra complications. As long as you use your common sense, these diagnoses can be used in the communication with your group of pupils equally well as the technical complications of the Smartboard, which are so much easier for your pupils to work out than for you.

Psychiatrists use the same strategy. 'Having diminished responsibility' is a response-dependent property that can be divided into several components. Some of these components can be attributed to the suspect, and some can be contributed to the circumstances. After all, sometimes the circumstances can be such that someone really cannot be held responsible for his behaviour. Suppose you stab your husband with a knife, after having been a victim of domestic violence for many years. Or think about a burglar who threatens you with a torch that looks like a gun to you. And it may be much more subtle and much more complex. Suppose you see a man walking with a child who is trying to wrestle free. You realise that you may well be witnessing a kidnapping; then you ask yourself what you should do, given that this might also be an everyday scenario of a father picking up his child from the playground because it is nearly dinnertime. Besides environmental factors that may make someone less accountable, a psychiatrist needs a component that he can attribute to the suspect: a component that makes this suspect – abstractly and in himself – susceptible to reason. It is this component that makes people accountable. But what kind of non-relational property could that be, and how can we calibrate it objectively, in the same way as we do with our weight in stones?

4. Out of sorts

Suddenly he is terribly annoyed by the ease with which he can flip the switch: from a talk with a prisoner to a meeting with the prison chaplain, to a talk with another prisoner, to a conversation with the warden, to a meeting with the probation department, to a phone call with his girlfriend, to a chat with his neighbour. And so on. It's a piece of cake to him; it all happens automatically. Always using the right tone: interested questions, engaged, open-minded, completely focused. He feels like a chameleon and he hates it.

He looks at his girlfriend, Janice. For years it has been Roger-and-Janice, as if it were one word, one item. And all these years he has been able to simply switch off; never before has his work been buzzing around in his head. A 15-minute walk is more than enough to lose his concerns about his clients. And just as easily his thoughts and ideas pop back into his head when he's back at work the next day. As a true professional he can routinely pick up exactly where they left off, with real attention for what is going on. He used to be proud of his ability to store and continue these conversations so easily. Sometimes he even needed to remind his client what they had been talking about previously. It is a talent; at least, that's what he has always thought until now.

Her name is Janice, too. Perhaps that is it? And he has heard terrible stories before, so why is this one having such an impact on him? Why can't he shake it off? She had lived in fear for such a long time, like in a tunnel that was becoming increasingly smaller. Violence having become such an ordinary thing in her home, she never knew what she had done wrong this time. She was always wary of the next outburst, knowing that it would come, just not *when* it would come. Humiliation and pain were commonplace. But perhaps he is so obsessed with the case because he really can't imagine what it must be like.

— What are you having, Roger?

Her voice hardly gets through to him.

— Roger? What are you having?

Janice looks at him from behind the menu, a questioning expression on her face. Only then does he notice that the waitress is standing next to their table, holding a pen and a notebook.

— O, erm, I'm sorry. I haven't really looked yet. I'll have the . . . the Thai omelette, please.

Vegetarian. Why on earth? What was the last time that he chose a vegetarian dish in a restaurant?

Janice looks at him. Sweet. Caring. She smiles.

— You're not yourself today, Roger. Thai omelette? I don't think you heard anything I said about Los Angeles, did you? It's not like you at all.

She looks at him.

— Is something the matter?

5. Fair, unbiased, neutral and impartial

You do not need any objectively determined legal competence if you are with your girlfriend in a restaurant and choose your main dish from the menu. You may be slightly distracted, you may have hardly looked at the menu, you may be a little out of sorts. And when you randomly choose a Thai omelette because you feel pressurised by the waitress, this choice is probably not well-considered. At least, this is what a person observing the scene behind a two-way mirror may be inclined to conclude. You haven't really seen what is on the menu and you probably haven't really thought about what you would like to eat. You are not even interested; but of course you have made this choice, and in the ensuing conversation you will take full responsibility for it. So later you will not send the waitress back because you feel it should have been abundantly clear to her that you are not the type of person who eats vegetarian meals. This is not the way you do things. This is not the way we do things.

Later, it becomes clear that you have not even heard your girlfriend's important and wonderful news – she has been asked to escort her aunt to Los Angeles – so now you have some explaining to do. Not every excuse will be good enough. But together, the two of you will be able to work it out. Perhaps there will be some damage, or some loss of face, but perhaps you will also gain some insight into your own good and bad patterns of behaviour. Perhaps there will be some bitterness, or disappointment, but perhaps there will also be some more intimacy because you have seen each other's weak spots and together you have built a safety net for these weak spots. This is the way you do things. This is the way *we* do things.

Suppose that all that time someone has been watching you from behind a two-way mirror. He saw your conversations with prisoners, with staff members, with the warden. He saw you walk home, phone your girlfriend, take the bus into town, order a delicious glass of dry, white wine, and he saw you take the menu. Perhaps he saw your eyes glaze over, saw you were preoccupied, noticed that you didn't hear what your girlfriend was telling you about Los Angeles, and saw the way you looked at the waitress's notebook. He also paid attention during the rest of the evening: he watched what you were doing, what you said, and tried to understand why you did what you did and why you said what you said. Suppose that he appeared from behind the mirror – just like your conscience used to bother you when as a child you were saying your prayers and thinking about the things that you had done wrong during the day. Would he simply know what had been going on, what you were made of, what mechanisms your behaviour was based on? Would he be able to pinpoint the reasons for your behaviour – like a neutral, objective scientist? Or would he ask you questions, would he want to know why you did what you did, why you failed to notice some things that were so obvious, why some realities didn't enter your awareness? Would he be like a friendly, attentive, perhaps slightly strict guardian angel, full of understanding but also critical, knowing that there were things that you should have known better and should have done better?

How much can you see behind a two-way mirror? And how well do you have to know someone, how intimately do you have to penetrate the deepest recesses of his soul, to be able to say – better, more correctly, and more convincingly than the person himself – that you know what it is that motivates him? And how much can you love somebody or hate somebody, if you explore the intimacy of their soul? Can you do this, behind a two-way mirror? And what would the effect of such a mirror be on *you*, if you were to sit behind it?

Of course, the two-way mirror is only a metaphor. I am using this metaphor to express that an expert witness needs to keep an appropriate distance from the people of whom he needs to ascertain whether or not they can be held fully accountable for their actions. What kind of distance is this? Of course we're not talking about physical distance. Just try to determine at a distance of 500 metres how heavy someone is or what hair colour he has. Not a chance. Nor are we talking about a cultural or social distance here: a senior citizen who has spent most of his life in a hamlet on Greenland does not have a better chance than your girlfriend of ascertaining whether or not you are out of sorts if you order a Thai omelette in a restaurant. So what kind of distance are we talking about? Emotional distance? Are we dealing with the distance that guarantees an impartial, unbiased and fair assessment of a suspect's accountability? It seems so. The mirror is supposed to filter out emotional involvement. It is to do with the distance necessary to avoid any distortion of the facts by the expert witness behind the mirror.

However, it is clear that we have an enormous problem here. 'Being accountable' is unmistakably a response-dependent property, a property in which two people come together: the person who is holding to account and the person who is held to account. You can try to break down this property into two components, and in the black-and-white areas this is no problem at all: a bottle of champagne cannot be held to account. It doesn't play along in our social interaction, and neither do guinea pigs, toddlers or people in a coma. You can try and hold them to account until you are blue in the face, but it is useless. An expert witness is only involved if we are stuck in a grey area, where it is really difficult to break down these components. In small-scale relationships we usually explore the grey areas together, to see whether we can solve them together. In these relationships – the domestic circle, friendships, small companies, neighbourhoods and so on – we try to sort things out, but not by breaking them down into several components or by looking at them from a distance, but rather by looking for each other's proximity.

This means that if you are unsure whether someone is still accountable, you pull out all the stops to make sure that you capture the other person's attention so that you can remind them of their responsibilities. This is a question of common sense, as discussed in Part 1: first, it is a question of adopting an investigative attitude to see what you can do with the grey; second, it is a question of trust and accommodation because you want to be able to appreciate the emotions of the people involved; and third, it is a question of respecting the entitlements and obligations that accompany your expectation that everybody will do their best to show understandable, efficient and good behaviour. Pulling out all the stops involves beautiful everyday phrases such as 'for better or for worse' and 'the easy way or the hard way'. This engagement is part of the deal; it is the heart of our common sense. You hold somebody to account 'the easy way or the hard way' because you think that he should be accountable. This implies your moral emotions, your justifiable indignation, your legitimate fear, your justified concern.

But how would you go about this as an expert witness? Would you observe from behind your two-way mirror how your client reacts to the attempts of the other people in the scenario to keep the suspect alert? And how would you separate the different components? The woman who has been abused and humiliated for years: how was she held to account by her husband? How can you see whether she can still listen to reason? And what reason are we talking about? Or will you, as an expert witness, appear from behind the mirror to take your client to task about her behaviour like a friendly, attentive and strict guardian angel? If so, how will you manage to divide this response-dependent property, this 'being accountable' that is hanging in the air between the two of you; how can you break it down into two components, one of which to be attributed objectively and impartially to your client.

This is my opinion and it doesn't bother me that I am simply taking a radical stance here. 'Accountability' is not an objectively determinable, response-independent property; it cannot be. 'Accountability' is a property that we have no choice but to attribute to each other in our attempts at humaning, in our attempts to live together with each other. Sometimes we give up. But this doesn't happen quickly. However, there are times when we just can't see a way out, times when a person just doesn't do any humaning. None at all. And this leaves a bitter taste.

At such a moment, it is weak and cowardly to resort to so-called objective facts that exclude ourselves and our engagement. The expert witness does not ascertain anything; he lets us know that it is a waste of our time, energy and attention to allow his client to play along with us at this moment. It is like in your childhood playground, when you sometimes didn't want a child to join in because he would ruin the game. It was harsh, very harsh.

But besides being harsh, it may also be mean if you are trying to hide behind a two-way mirror, behind a so-called truth that you have determined in a fair, unbiased, impartial and objective manner. It is not us, it is her. And then actually it is not her because she can't really help it, because she is not accountable. So that in the end it is us after all, because we actually do not know how we could held her to account. And in this way we accept that she is not one of us. She is not humaning; at least, not like us.

6. Now what?

By now there are about 7.5 billion of us on our green planet: this is a huge number of people. Of all these people, I may know a few hundred, perhaps a thousand if you use 'knowing' in the broadest sense of the word. Of these acquaintances, I know ten, twenty, perhaps thirty well, and one or two very well. Just imagine how many strangers there are on earth. There are so many of you! Still, I will get to hear if something goes wrong, even if it is something minor. And it won't just be me. Every one will hear it. Just as you will get to hear if something minor happens to me. In that respect it is just as if we are all living together in some dead-end street, with the neighbours watching everything that happens.

However, this is a misleading image. Although the mass media have found their place in even the smallest corners of our society, their reporting is extraordinarily fragmented, broad, one-sided and cursory. The neighbours have very specific interests and a low attention span. They want to know in a fraction of a second what reasons someone may have had for a completely incomprehensible act, a crime that has shaken a local community and that has unhinged the lives of many people. They only allow the suspect a fraction of a second to justify himself. And the rest of us, the 7.5 billion people in the grips of mass media, only get a fraction of a second to ask the suspect what has gone wrong, at some place, at some time; what has gone so drastically wrong that she now has to live with the consequences of her incomprehensible act.

Where do you start when you have so little time? How could this happen? Who is responsible? Which authority has . . .

Click.

Next subject. The demolition of the old Town Hall needs to be delayed because an endangered barn owl is nesting in the chimney. Experts are currently discussing whether . . .

Click.

It is not easy to find the human dimension in a world that seems on the one hand as accessible and transparent as a dead-end street in our village, but on the other hand as anonymous as a far-away metropolis. Where do you begin if there are 7.5 billion people? You can't even shake everybody's hand; and shaking hands, what is the use of that? How can you, together with so many people, reach the conclusion that one individual has diminished responsibility whereas another person can definitely be held accountable for his deeds? How do you approach something like that? How do we do this with other response-dependent properties, such as convincing, funny, exciting, decisive, disgusting, rich and intelligent? It is an endless list, but you may already have noticed a great variety in this short version, and you may have realised that it is unlikely that there is one common or ideal way of addressing all these properties.

Suppose that we, all 7.5 billion of us, have indeed been given the task to reach a unanimous decision whether someone has a specific response-dependent property. Is Benedict Cumberbatch really as good an actor as he thinks he is? And is *The Silence of the Lambs* really so exciting? Bill Gates so rich? And Stephen Hawking so intelligent? Can you see the diversity? I can imagine that we would appeal to a reputable jury, for example if we were talking about 'acting', it would be the jury of the BAFTA awards. And perhaps we could also use a jury if we're talking about 'exciting', but then I would like to have a jury that is not completely enamoured with the special effects that you see in most movies nowadays. You can probably see where this is going. 'Exciting' may well be mostly a question of taste, and we might lose some interesting and relevant differences if we were to insist that we reach a unanimous verdict. Perhaps you think the same is true for acting, and I can see your point, as you may be really impressed with all of the cast in *EastEnders* – unlike me. With regard to these properties, unanimity may lead to impoverishment, to the neglect of relevant differences.

Do we have any reason to expect unanimity when we are talking about 'rich' or 'intelligent'? And what would that mean? Do we need a jury for these properties? You would not think so. After all, 'rich' is only a matter of counting money. But appearances can be deceptive: undoubtedly you will need a group of top accountants to determine what belongs to Bill Gates's assets and how much this is worth. Are these accountants a jury? If so, on what do they base their judgement? You run into comparable problems if you want to assess intelligence. An IQ test may sound like an objective measuring device, like some kind of gauge. But if you know how these tests are put together and how they are validated, and how you can train to achieve a better result, then you know better than that. It may well be a good idea to measure your intelligence in a talent show, with a jury that can give their verdict, with a live studio audience that can applaud, and with people at home who can vote by text message.

I realise full well that these are controversial ideas, but this is only fuel to my fire, and only reinforces the thesis that I defended in Chapter 8: that response-dependent properties show us that we only have a common world as a language community. And that this language community is built on the idea that we share a normal language ability and find ourselves in favourable circumstances. In this common world we have a great deal of response-dependent properties because we do our humaning together; these properties are strung between human beings because of the way in which we react to one another. You can either agree on these properties, or you can debate them. These are two ways in which to react to one another, but there are many different ways. We can vote for each other in talent shows, evaluate each other, demolish each other or instead extol one another's virtues, and so on. As a result of these different ways of reacting, we have these properties in the only way possible: *together*, by forming one language community, by ascribing a normal language ability to one another and by believing that we find ourselves in favourable circumstances. And just as in the *X-factor*, the jury and the audience together decide that they have a normal language ability and that the circumstances are favourable, and they show this by together reaching the conclusion that Paul Potts and Susan Boyle are excellent singers.

That is how we deal with response-dependent properties. *Together.* That is also how we deal with that crucial, fundamental property, the property that can bind us together and split us asunder, the property that distinguishes between humaning and not humaning: accountability. And because we have become so used to scientists and their missing information subroutines, we are happy to leave the tasks that overwhelm us to expert witnesses. For example, we don't know whether to condemn another human being for not doing his humaning properly. Thus, we gladly leave the task to a psychiatrist, who will then be asked to determine whether or not the suspect has diminished responsibility. As if it is written on the suspect's forehead in a secret language or can be found in his DNA.

I do not envy these expert witnesses who we have more or less condemned to answer a question on our behalf – in a fraction of second – that none of us understands well enough by a long shot. I admire their courage. I would like to encourage them, and this is why I argue in this book not only for a reappraisal of our common sense and for an investigative attitude, but also against the complacency of lay people who expect that for every question there is a missing information subroutine in science.

I would like to finish with three short remarks.

First, we are well aware that the X-factor is all about the moment. Of course Paul Potts and Susan Boyle are excellent singers, but one bad performance would have been enough to prematurely oust them. Nobody can claim favourable circumstances forever, and this gives response-dependent properties a temporary and situational character. If I have had a good night's sleep and have prepared well I can explain complicated things really very clearly. Still, I also know there are plenty of students who think I can be rather vague.

Second, it is only in very mysterious circumstances and if someone's actions are really incomprehensible that questions are raised about this person's accountability. Such questions only occur in scenarios in which it is abundantly clear that we are dealing with a grey area. As I have argued in Chapter 5, this requires an investigative attitude and as such also a great willingness and commitment to reach out to the other person, to find a way to address him, to hold him accountable 'the easy way or the hard way'. That is why in such scenarios it is hardly possible to break down a response-dependent property such as accountability into objective and subjective components.

Third, if you do not think your fellow human is accountable, you switch to a different register to address him. On the one hand, you do so because this person with diminished responsibility will also address you in a totally different register – for example as a raving lunatic trying to attack you with a knife. On the other hand, you do so because you are no longer simply holding the person with diminished responsibility to account as a criminal, but you're addressing him as a dangerous, insane patient who needs treatment rather than punishment.

In other words, we will never stop addressing one another; this will go on and on, in whatever way. Holding one another accountable is what we do, this is what humaning comes down to. We can use all the intellectual help science can offer us. But we cannot leave our humaning to experts. We will have to do it ourselves. Using our common sense.

Note

1   I have learned much about this language from Wallace (1994) and Vargas (2013).

Epilogue

I enjoyed writing this book enormously. The intensity of this project led to hours of total concentration, to days rolling into one when the deadline came nearer, to dreams at night about passages in which I got lost, to walks to help me organise my thoughts, to frustratingly rearranging paragraphs that would not flow, to a garden that has been neglected, to images that suddenly presented themselves to me, to surprisingly lucid reasoning that looked at me benignly from the computer screen, to words that made my face light up in agreement and satisfaction – all of these were experiences that I enjoyed immensely, and that made writing this book a delightful occupation.

However, despite all the hours at my computer it hasn't been a lonely adventure. Not at all. The first ideas were formulated in Greece, during an academic retreat in Artisa in November 2011. I returned there twice in April and October 2012. When I was there, not only did I work hard, but I also enjoyed the inspirational atmosphere, the glorious weather, the good food, the welcoming company and the constructive feedback to my texts. Thank you, Louise Thoonen and Celeste Neelen. I can heartily recommend going on an *Academic and Art Retreat* on the Greek coast.

Many other people have read my texts and given me important, critical and constructive feedback. First of all, Wim de Muijnck and Melissa van Amerongen come to mind. Just like with my previous book for a general audience, *Become a philosopher yourself*[1], they read the entire manuscript and suggested dozens of improvements. I would like to thank them wholeheartedly, especially because whenever I provided them with new text at the very last minute, they always found time to give a prompt response. Moreover, I am touched by the fact that my wife Liesbeth Stöfsel read the whole manuscript with a great deal of affection, that she made perceptive and helpful suggestions, and that she makes it so easy for me to do my humaning. Finally, Elianne Muller also carefully went through the complete text and her suggestions regarding content as well as her linguistic and editorial advice helped me tremendously. Thank you.

I also asked a number of people to read specific chapters. I would also like to thank them for their comments which were sometimes concise and sometimes very extensive: Anna Bosman, Rob van Gerwen, Clemens Raming, Derek Strijbos, Femke Takes and Rolf van Til.

Next, I would like to thank Edo Klement for his faith in this project, for the apt title that he suggested, for the well-organised publishing company that he represents, and for the 'i's that he dotted and the 't's that he crossed in my text.

There is a further line needed to this epilogue, now that this English translation appears in print: thank you very much Fulco Teunissen and Kate Kirwin of *Twelvetrees Translations* for the splendid translation! It's a wonderful experience to recognize my own voice in this foreign language.

Finally, I would like to use this opportunity to thank two groups of people who have not been involved in the shaping of this book and who probably have never realised how much their efforts and their ambitions have been an inspiration to me. I am referring to my colleagues and students at the Behavioural Science Institute of Radboud University, Nijmegen. Although I am greatly concerned about the scientification of our everyday life, although I want to revolt against the societal arrangements that systematically reward our lazy brains, and although I am very critical about the direction in which the behavioural sciences are heading these days, in this book I have not tried to make a case against science. I love the intellectual workmanship, the methodological care and the critical eye for detail that can still be found all over academia. I love the perception, thoroughness and engagement of my colleagues and students. The behavioural sciences are close to my heart; just as is our common sense.

I hope my students will have the opportunity to climb on the shoulders of my colleagues and then notice the abyss that they will have to avoid. I hope they will have a wonderful future in which they can contribute, from the inside, to the fight against the non-productive assumption that there is an antagonism between the behavioural sciences and common sense. There is no truth in that: the behavioural sciences are an extrapolation of our folk psychology, just as the natural sciences are an extrapolation of our folk physics. The natural sciences can improve our insight into the causal mechanisms that we use in our expectations about objects. Similarly, the behavioural sciences can improve our insight into the normative patterns that we use in our expectations about our fellow human beings. But we shouldn't make the mistake of thinking that normative patterns, like causal mechanisms, can be best studied from a neutral, objective and impartial perspective.

The latter observation has great consequences, both for the behavioural sciences and for us, human beings, who want to use it to our benefit. For the behavioural sciences it means the continuation of the so-called *Methodenstreit* – we still do not know *how* science could or should be conducted.[2] For us, human beings, it means making a case for our common sense, a case for the investigative attitude and a case *against* the habit of thoughtlessly passing on all our questions to the experts and their missing information subroutines.

I hope that my students and my colleagues have the opportunity to continue their ambitious work towards filling our toolbox with all kinds of behavioural-scientific theories and observations. I especially hope that they have the opportunity and the time to look extensively, with attention and surprise at their own hands, the hands that will use the tools in the toolbox; in other words: their common sense.

I want to thank them for the insight that they have granted me: there is nothing wrong with their hands!

Notes

1   J. Bransen, *Word zelf filosoof*. Amsterdam: Veen Magazines, 2010. This book is only available in Dutch.
2   This is beyond the scope of the present book. I once wrote a lemma on this topic: Bransen (2001b).

Bibliography

Aristotle (1930) *Ethica Nicomachea. The Nicomachean Ethics of Aristotle*, translated by Sir David Ross. Oxford: Oxford University Press.

Baier, A.C. (1986) Trust and antitrust. *Ethics*, 96, pp. 231–60.

Baker, L.R. (1999) What is this thing called 'commonsense psychology'? *Philosophical Explorations*, 2(1), pp. 3–19.

Blackburn, S. (1998) *Ruling Passions*. Oxford: Oxford University Press.

Blumer, H. (1969) *Symbolic Interactionism: Perspective and Method*. Englewood Cliffs: Prentice-Hall.

Brandom, R. (2000) *Articulating Reasons: An Introduction to Inferentialism*. Cambridge/ Mass.: Harvard University Press.

Bransen, J. (2001a) On exploring normative constraints in new situations. *Inquiry*, 44(1), pp. 43–62.

Bransen, J. (2001b) Verstehen and Erklären; The Philosophy of. In N. Smelser and P. Baltes (eds.), *International Encyclopedia of the Social and Behavioral Sciences*. Oxford: Elsevier Science Ltd., pp. 16165–70.

Bransen, J. (2008) Personal Identity Management. In C. Mackenzie and K. Atkins (eds.), *Practical Identity and Narrative Agency*. New York: Routledge, pp. 101–20.

Bransen, J. (2010) *Word zelf filosoof*. Diemen: Veen Magazines.

Bransen, J. (2014) Loving a Stranger. In C. Maurer, T. Milligan and K. Pacovska (eds.), *Love and Its Objects: What Can We Care For?* Basingstoke: Palgrave Macmillan, pp. 143–59.

Bratman, M.E. (2014) *Shared Agency: A Planning Theory of Acting Together*. Oxford University Press.

Casati, R. and Tappolet, C. (eds) (1998) *Response-Dependence: European Review of Philosophy*. Stanford: CSLI Publications.

Cassirer, E. (1923–9) *Philosophie der symbolischen Formen*. 3 volumes. Berlin: Bruno Cassirer Verlag.

Damasio, A. (1994) *Descartes' Error*. New York: Putnam.

Davidson, D. (1985) Incoherence and irrationality. *Dialectica*, 39(4), pp. 345–54.

Dennett, D. (1987) *The Intentional Stance*. Cambridge/Mass.: M.I.T. Press.

Dewey, J. (1922) *Human Nature and Conduct: An Introduction to Social Psychology*. New York: Henry Holt & Co.

Dewey, J. (1941) Propositions, warranted assertibility, and truth. *Journal of Philosophy*, 38, pp. 169–86.

Feinberg, J. (1980) The Child's Right to an Open Future. In W. Aitken and H. LaFollette (eds.), *Whose Child? Children's Rights, Parental Authority, and State Power.* Totowa, NJ: Rowman & Littlefield., pp. 124–53.

Fonagy, P., Gergely, G., Jurist, E. and Target, M. (2004) *Affect Regulation, Mentalization and the Development of the Self.* New York: Other Press.

Foucault, M. (1969/2002) *The Archeology of Knowledge*, translated by A.M. Sheridan Smith. London: Routledge.

Frankfurt, H.G. (2006) *Taking Ourselves Seriously & Getting It Right.* Stanford: Stanford University Press.

Garfinkel, H. (1967) *Studies in Ethnomethodology.* Englewood Cliffs: Prentice Hall.

Gehlen, A. (1940) *Der Mensch, seine Natur und seine Stellung in der Welt.* Berlin: Junker und Dünnhaupt.

Gehlen, A. (1957) *Die Seele im technischen Zeitalter: Sozialpsychologische Probleme in der industriellen Gesellschaft.* Reinbek: Rohwolt.

Gibson, J.J. (1979) *The Ecological Approach to Visual Perception.* Boston: Houghton Mifflin.

Goffman, E. (1956) *The Presentation of Self in Everyday Life.* Edinburgh: University of Edinburgh Social Sciences Research Centre.

Goodman, N. (1978) *Ways of Worldmaking.* New York: Hackett Publishing.

Haack, S. (2009) *Evidence and Inquiry: A Pragmatist Reconstruction in Epistemology.* New York: Prometheus Books.

Haack, S. (2013) *Putting Philosophy to Work: Inquiry and Its Place in Culture.* New York: Promotheus Books.

Hacking, I. (1975) *Why Does Language Matter to Philosophy?* Cambridge: Cambridge University Press.

Hacking, I. (2002) *Historical Ontology.* Cambridge/Mass.: Harvard University Press.

Hacking, I. (2006) Making up people. *London Review of Books*, 28(16), pp. 23–26.

Haybron, D.M. (2008) *The Pursuit of Unhappiness: The Elusive Psychology of Well-Being.* Oxford: Oxford University Press.

Heidegger, M. (1953/2000) *Introduction to Metaphysics*, translated by Gregory Fried and Richard Polt. New Haven: Yale University Press.

Hieronymi, P. (2008) The reasons of trust. *Australasian Journal of Philosophy*, 86(2), pp. 213–36.

Horn, M. (2004) A rude awakening: what to do with the sleepwalking defense? *Boston College Law Review*, 46, pp. 149–82.

Kahnemann, D. (2011) *Thinking, Fast and Slow.* London: Penguin Books.

Kuhn, T.S. (1962) *The Structure of Scientific Revolutions.* Chicago: The University of Chicago Press.

Macdonald, G. and Pettit, P. (1981) *Semantics and Social Science.* London: Routledge & Kegan Paul.

Maturana, H. and Varela, F. (1998) *The Tree of Knowledge: The Biological Roots of Human Understanding.* Boston: Shambhala.

McDowell, J. (1994) *Mind and World.* Cambridge/Mass.: Harvard University Press.

McGeer, V. (2002) Developing trust. *Philosophical Explorations*, 5(1), pp. 21–38.

Mead, G.H. (1934) *Mind, Self & Society: From the Standpoint of a Social Behaviorist*. Chicago: Chicago University Press.

Morton, A. (2003) *The Importance of Being Understood: Folk Psychology as Ethics*. London: Routledge.

Parsons, K. (ed.) (2003) *The Science Wars: Debating Scientific Knowledge and Technology*. New York: Prometheus Books.

Pettit, P. (1991) Realism and response-dependence. *Mind*, 100(4), pp. 587–626.

Prinz, J. (2004) Emotions Embodied. In R. Solomon (ed), *Thinking About Feeling: Contemporary Philosophers on Emotions*. Oxford University Press.

Quine, W.V. (1957) The scope and language of science. *British Journal of the Philosophy of Science*, 8, pp. 1–17.

Rawls, J. (1971) *A Theory of Justice*. Cambridge/Mass.: Harvard University Press.

Rosen, S. (1979) *The Limits of Analysis*. New Haven: Yale University Press.

Schutz, A. (1967) *The Phenomenology of the Social World*, trans. G.Walsh and F. Lehnert, Evanston: Northwestern University Press.

Sterelny, K. (2003) *Thought in a Hostile World: The Evolution of Human Cognition*. Oxford: Blackwell.

Stroebe, W., Postmes, T. and Spears, R. (2012) Scientific misconduct and the myth of self-correction in science. *Perspectives on Psychological Science*, 7(6), pp. 670–88.

Taylor, C. (1985) *Human Agency and Language: Philosophical Papers I*. Cambridge: Cambridge University Press.

Tiberius, V. (2008) *The Reflective Life: Living Wisely With Our Limits*. Oxford: Oxford University Press.

Tinbergen, N. and van Iersel, J.J.A. (1947) "Displacement reactions" in the three-spined stickleback. *Behaviour*, 1(1), pp. 56–63.

Vargas, M. (2013) *Building Better Beings: A Theory of Moral Responsibility*. Oxford: Oxford University Press.

Velleman, J.D. (2003) Narrative explanation. *The Philosophical Review*, 112(1), pp. 1–25.

Velleman, J. D. (2009) *How We Get Along*. Cambridge: Cambridge University Press.

Wallace, R.J. (1994) *Responsibility and the Moral Sentiments*. Cambridge/Mass.: Harvard University Press.

Watts, A. (1951) *The Wisdom of Insecurity: A Message for an Age of Anxiety*. San Francisco: Pantheon Books.

Williams, B. (1978) *Descartes: The Project of Pure Enquiry*. Harmondsworth: Penguin.

Williams, B. (2002) *Truth and Truthfulness: An Essay on Genealogy*. Cambridge/Mass.: Harvard University Press.

Wittgenstein, L. (1953) *Philosophical Investigations*. translation: G.E.M. Anscombe. Oxford: Blackwell.

Wittgenstein, L. (1977) *Remarks on Colour*. G. Anscombe (ed.). Oxford: Blackwell.

Index

n refers to note